T0332671

Machine Translation and Transliteration Involving Related and Low-resource Languages

Machine Translation and Transliteration Involving Related and Low-resource Languages

Anoop Kunchukuttan
[Microsoft, India]
Pushpak Bhattacharyya
[IIT Bombay- Maharashtra, India]

CRC Press is an imprint of the
Taylor & Francis Group, an **informa** business
A CHAPMAN & HALL BOOK

First edition published 2022
by CRC Press
6000 Broken Sound Parkway NW, Suite 300, Boca Raton, FL 33487-2742

and by CRC Press
2 Park Square, Milton Park, Abingdon, Oxon, OX14 4RN

ISBN: [978-0-367-56199-4] (hbk)
ISBN: [978-0-367-56200-7] (ebk)
ISBN: [978-1-003-09677-1] (pbk)

Typeset in Times New Roman font
by KnowledgeWorks Global Ltd.

Contents

Preface

The book is about machine translation and transliteration of related languages. Information transfer between related languages is a real need, since people speaking these languages often reside in close proximity (like, adjacent states/provinces) and require communication for social, business and administrative purposes (for example, land disputes). Script, lexicon, syntax, semantics and pragmatic barriers interfere with mutual understanding between two people communicating in their own languages. For example, a native speaker of Assamese from a remote village in Assam, will hardly comprehend anything in Tamil. However, she will understand quite a lot of Bengali, since Assamese and Bengali are "related" languages. Languages are said to be "related", when they are descended from the same parent (*e.g.*, Bengali and Assamese, both descendants of Sanskrit) or are in close proximity (*e.g.*, Marathi and Kannada, Maharashtra and Karnataka are bordering states). Related languages have high proportions of shared phonology, morphology, vocabulary, proverbs and idioms and socio-cultural expressions.

Many times, therefore, facilitating mutual intelligibility between two related languages boils down to simply arranging for crossing the script barrier, *i.e.*, providing **transliteration**. Punjabi speakers capable of reading only Gurumukhi script will understand a lot of Hindi content written in Devanagari if the Devanagari text is transliterated into Gurumukhi. Transliteration is the task of transforming from one script to another, preserving the sound. We can call this task *script transfer*. About half of the current book is devoted to the discussion of challenges in automatic transliteration and solutions therein.

Like all Natural Language Processing (NLP) tasks, transliteration too has to grapple with *ambiguity*. The ambiguity arises from the one-to-many mapping of graphemes of source text to that of the target. Thus the surname of a famous Indian actor "Amir Khan" when transliterated from Roman to Devanagari is written as खान (khaana), while the transliteration of "khanij" (mineral) is खनिज (khanija), meaning *born in the 'khana' or mine*). Thus the same vowel "a" in roman transforms to "aa" or "a" sound. This is the ambiguity challenge that needs to be resolved. It is very difficult to construct *rules* for such disambiguation. The solution is predominantly data driven. Machine learning models are trained with parallel transliteration strings, which acquire pattern mappings in a context and apply them in test situations.

For related pair of languages, script transfer is much less beset with ambiguity than in case of distant pairs. The sound mapping is almost one to one.

The next set of barriers after script are vocabulary, morphology, syntax (word order), idioms and cultural expressions. Related languages share these entities in large proportion. For example, categories of words called "tatsam" — meaning "identical form in sanskrit" — and "tadbhav" — meaning 'tatsam derivative' are shared quite a lot across Indo Aryan group of languages (*e.g.*, "chandra", "chaand", "chandaa" — all of them meaning 'moon' appear in all North Indian languages). Morphological isomorphism too abounds in pairs of related languages. For example, the accusative, "raam ko" (to raam) in Hindi maps to "raam ke" in Bengali, which is one-to-one. Here crossing the language barrier involves "morphology transfer" and is almost always one-to-one for related languages. The case marker appears mostly as a postposition or suffix with very little ambiguity.

Proverbs and idioms that epitomise time tested learnings from everyday observations and experiences are often shared literally between pairs of related languages. For example, the oft used expression for an incompetent worker in Hindi is "naach naa jaane aangan tedaa" (literally — "a dancer who does not know how to dance says the floor is uneven") has identical expression in Bengali, Marathi, Punjabi, and Gujarati which are all related languages. Proverb and idiom translation is a very difficult task, needing simply memorization of the mappings. Hence, the transfer is much easier for related languages.

Thus, translation and transliteration of related languages is a great practical requirement and developing solutions for these problems can take advantage of the similarities between related languages that is alluded to in the previous paragraphs. While language relatedness is an advantage, many related languages have few linguistics resources to develop practical solution. The central thesis of this book is that utilizing language relatedness can help bridge the disadvantages due to the limited availability of linguistics resources.

Computationally speaking, the techniques for translation and transliteration of related languages — as described in the current book — make use of two basic ideas: (1) subword representation and (2) multilinguality.

Subword representation refers to breaking words into their smaller units. The treatise herein picks up two kinds of subwords: *orthographic syllables (OS)* and *byte pair encodings (BPEs)*. OSs are word segments starting with a consonant and ending in the first encountered vowel reading left to right. Thus the OSs in the word 'little" are "li" and "ttle'. BPEs are statistically significant (count-wise) segments in large corpora. Thus BPEs in "quickly" are "quick" and "ly", both of which are likely to be very frequent strings in large corpora of English. We show that designing the right subword units for translation can best utilize the vocabulary overlap between related languages and improve translation quality with limited parallel corpora.

The second basic idea, multilinguality, essentially brings together training data for multiple languages. For related languages, simple methods can be used to effectively pool data from different languages and derive benefits from transfer of knowledge across languages. We show that such multilingual solutions can improve translation and transliteration quality for related languages.

Organization of the monograph

The first two chapters are introductory chapters. Chapter 1 introduces the notion of related languages, sources of language relatedness and key properties. We motivate the utilization of language relatedness for translation and transliteration and provide a high-level summary of the research contributions presented in this monograph. In Chapter 2, we present an overview of the past work on machine translation involving related languages. The rest of the monograph is divided into two parts: Part I and Part II deal with machine translation and machine transliteration involving related languages, respectively.

Chapters 3 to 6 constitute Part I. In Chapter 3, we propose methods to improve SMT between related languages using subword-level translation units. Chapter 4 continues our investigations on subword-level SMT and explores methods for improving the performance of subword-level SMT. Chapter 5 explores how data from multiple languages can help improve SMT between related languages. Chapter 6 presents an extensive case study of SMT involving major languages from an important group of related languages: languages spoken in the Indian subcontinent.

Chapters 7 and 8 constitute Part II. The focus of these chapters is the improvement of machine transliteration by utilizing multilinguality as well as script similarity between related languages. In Chapter 7, we propose a method for unsupervised transliteration between related languages which utilizes script similarity to inject inductive bias in the learning process. In Chapter 8, we propose a multilingual neural transliteration model for transliteration involving related languages.

Finally, Chapter 9 provides a summary of the work presented in the monograph. It also presents the major conclusions and findings from our research as well as discusses future directions of work.

Acknowledgments

Anoop would like to thank his parents, P. Kunchukuttan and Santha Kunchukuttan for their support and encouragement to follow his interests. He would also like to thank his brother, Anand, who has handled family responsibilities — allowing Anoop to dedicate himself to his research. He also thanks his cousins, Sujith and Sunil, uncle and aunt for their guidance in his formative years. He expresses his gratitude to wife, Sangeeta, for her constant support during the writing of the book. Finally, he would also like to thank his advisor and co-author Pushpak Bhattacharyya who introduced him to NLP and has been a guide and inspiration. The wonderful team he has built at the Center for Indian Language Technology (CFILT) has been a force multiplier for his learning. He would also like to thank his colleagues and collaborators who have been a great support during his time at IIT Bombay as a grad student: Abhijit Mishra, Joe Cheri Ross, Aditya Joshi, Brijesh Bhatt, Ajay Nagesh, Balamurali A. R., Manish Shrivastava, Manoj Chinnakotla, Rudramurthy, Girish Ponkiya, Kevin Patel, Sandeep Mathias, Arjun Atreya, Sudha Bhingardive, Raksha Sharma, Vasudevan, Diptesh Kanojia. The rich discussions and interactions with them on myriad topics have enriched his work and helped him grow.

Pushpak would like to thank his grandfather, the late Purna Chandra Sastri, who was a renowned scholar of Sanskrit, for introducing to him the ancient and beautiful language, and his father, the late Paramesh Chandra Bhattacharjee, and English teacher Shri Atul Pal, for making him aware of the richness and beauty of English language. His mother has been a constant inspiration for being innovative and original. His brother Pallab has shouldered many family responsibilities, leaving him time for demanding academic work. Pushpak's wife Aparna and son Raunak have provided the love and care at home that sustains and nourishes. Finally, Pushpak puts on record his deep gratitude to Dr. Vineet Chaitanya and Prof. Rajeev Sangal for training him in NLP, and his many collaborators in NLP — Prof. Dipti Mishra Sharma, Prof. Christian Boitet, Prof. Igor Boguslavsky, Mr. Hiroshi Uchida, Dr. Asif Ekbal, Dr. Sriparna Saha to name only a few — for the NLP journey. This NLP journey also would have been impossible without the intimate interaction with many great students he has had in CFILT including his coauthor in the current book, Anoop. The list of students is too long to enumerate; some of the names appear in the list of fellow students Anoop has mentioned above. Pushpak too acknowledges their intellectual companionship and many faceted help. And he adds to this list the names of Mitesh Khapra, Ritesh Shah, Ganesh Ramakrishnan, Jyotsna Khatri, Tamali Banerjee, Manasi Kulkarni, Brijesh Bhat, Sachin Pawar, Sapan Shah and many others. I am omitting the masters and bachelor degree students, because that list is really long!

We would like to thank all our collaborators whose inputs have made this research rich: Rajen Chatterjee, Abhijit Mishra, Mitesh Khapra, Ratish Puduppully, Ritesh Shah, Raj Dabre, Piyush Dungarwal Rohit More, Gurneet Singh, Maulik Shah, Pradyot Prakash, Palak Jain, Pratik Mehta, Sandhya Singh, Ritesh Panjwani. Thanks to Rajita Shukla, Jaya Jha, Laxmi Shukla, Gajanan Rane and many members of the Indo-Wordnet and ILCI teams across India whose efforts to create linguistic resources provided us the basic datasets for our work. We would like to thank Aditya Joshi, Rudramurthy and Sandeep Mathias who went read Anoop's thesis and gave their valuable feedback.

We would like to thank the Indian Institute of Technology Bombay and the Department of Computer Science and Engineering for their support over the years. We would also like to thank the administrative staff at CFILT for their support and making CFILT a great place to work. We would like to thank various agencies of the Government of India which have funded the research at CFILT over the years: Department of Science and Technology, Ministry of Electronics and Information Technology and Ministry of Human Resource Development. Anoop would also like to thank Xerox Research Center India (now Conduent Labs India) who supported his research.

Anoop Kunchukuttan
Pushpak Bhattacharyya

List of Figures

List of Tables

Chapter 1

Introduction

Language is one of the most remarkable and uniquely human abilities. *Natural language is a complex and versatile tool for communication which differentiates* `homo sapiens` *from other species*. The manipulation of symbols allows representation of objects, emotions, motivations, abstract thought, *etc.* Other species communicate in less complex ways. It is widely believed by linguists that full language capacity had evolved by 100,000 B.C.E.[1]. Language is undoubtedly one of the major factors in the rapid progress of humans since it allows faster dissemination of information/knowledge than what evolution and genetic mutation would permit. It enables diverse modes of social organization allowing humans to co-operate for achieving complex goals. As human societies grew and became ever more complex, humans invented writing around the 4th century B.C.E. to maintain and organize information. Writing enabled representation and recording of language, making long-distance and long-term dissemination of information and thought possible. It also enabled analysis and manipulation of large amounts of knowledge and information.

Ironically, *language can also act as a barrier to communication*. Language is not static, it is a system of shared conventions that changes over time. Over time, modern humans first spread all over Africa by around 150,000 B.C.E and then stepped out of Africa around 70,000 years ago. As humans spread out over the entire planet, different communities got segregated and language evolved independently in each of these communities. It led to the creation of multiple languages resulting in the great diversity we see in human languages. A vast majority of these languages are not mutually intelligible with each other. Languages also became repositories of culture, heritage and identity, something that persists to this day. A variety of writing systems also evolved making it non-trivial to understand knowledge recorded in multiple writing systems.

We know that, at least in recorded history, different cultures have felt the need to develop mechanisms to overcome language barriers and establish lines of communication. Thus arose the need for translation of languages and transliteration of written text from one script to another. One definition of translation is[2]:

Definition 1.0.1. Translation is the communication of the meaning of a source-language text by means of an equivalent target-language text.

[1] `https://blog.linguistlist.org/ll-main/ask-a-linguist-how-old-is-language`
[2] `https://en.wikipedia.org/wiki/Translation`

Traditionally, translation was achieved using human translators who would master more than one language. In addition, most cross-lingual communication was mediated through some languages which served as *lingua franca*[3] *e.g.,* Latin and Greek in ancient Europe, Sanskrit in ancient India, Arabic in the Middle East and English at a global level in modern times. Thus, translation played a major role throughout history in connecting peoples and cultures. Given the expertise required in mastering the nuances of multiple languages, the benefits of translation would have been limited to only a section of society who needed these services for the conduct of their professions. Manual translation is not scalable.

1.1 Need for Machine Translation and Transliteration

The modern age heralded the industrial and digital revolutions which transformed means of transportation and communication. We can travel to different parts of the globe in a short time. Information and knowledge from across the globe are available at our fingertips, especially with the advent of the Internet. We can communicate with people across the globe instantaneously. Hence, we have seen an explosion in our communications for administrative, business and cultural purposes. The world is highly interconnected and our interactions have global manifestations and implications.

While advances in transportation and communication technologies have reduced the physical barriers to communication, barriers due to linguistic divergences poses greater challenges. Given the degree of interconnectedness and resulting human communication needs, manual translation is no longer scalable to satisfy these requirements. Hence we need methods to automate translation of natural language *i.e., machine translation* (MT).

Different paradigms of machine translation have been proposed in the previous 60 years or so since investigations into machine translation began. In the earlier days, **rule-based machine translation** (Hutchins and Somers, 1992) was the dominant paradigm. This system relied on experts writing intricate and exhaustive rules based on deep understanding of language structure and language divergence. With the increased availability of translated data *viz.* parallel corpora, empirical approaches to translation (*e.g.,* **Statistical Machine Translation** (SMT) (Koehn et al., 2003) and **Neural Machine Translation** (NMT) (Sutskever et al., 2014; Bahdanau et al., 2015)) which try to automate discovery of translation patterns (word translations, phrase translation, translation rules *etc.*) were extensively explored. These empirical, data-oriented methods are the state-of-the-art methods and most research in translation has gravitated to such methods. The principal drivers of this shift are: (i) less dependence on expensive and scarce linguistic expertise, (ii) robustness to the diversity

[3] A language that is adopted as a common language between speakers whose native languages are different (Oxford Dictionary).

of language phenomenon and noisy input, (iii) ease of maintenance and (iv) rapid prototyping and development.

Given this evolutionary trend and its advantages, the work described in this monograph is based on these empirical methods, but draws upon linguistic knowledge to make learning more resource-efficient.

Sometimes, we need to convert text in one script to another script *i.e., machine transliteration*. Transliteration is particularly required for reading proper names written in one script in another script. It is a useful component of machine translation, cross-lingual information retrieval and similar multilingual tasks. Li et al. (2009) define transliteration as:

Definition 1.1.1. Transliteration is the conversion of a given name in the source language (a text string in the source writing system or orthography) to a name in the target language (another text string in the target writing system or orthography), such that the target language name is: (i) phonemically equivalent to the source name, (ii) conforms to the phonology of the target language and (iii) matches the user intuition of the equivalent of the source language name in the target language.

1.2 Need for Machine Translation involving Related Languages

From a practical standpoint, the demand for translation services is not uniform across all language pairs. There is little demand for translation among many language pairs. For instance, there is little interest in translation between Hindi[4] and Hausa[5]. Building good translation systems needs investment in creation/collection of parallel corpora and other linguistic resources as well as linguistic and machine learning expertise. It would, therefore, be prudent to focus on languages which need translation services.

One major use-case for translation services arises among people living in contiguous areas, speaking related languages. These people have cultural and economic ties and communicate heavily amongst themselves for administrative, business and social needs. For instance, the European Union (EU) is home to around 700 million[6] people speaking a wide variety of Indo-European languages with deep economic ties and substantial political and cultural interactions. Another example is the Indian subcontinent, whose 1.5 billion[7] people predominantly speak various Indo-Aryan and Dravidian languages. Hence, translation services for these languages is an important requirement. Two translation scenarios involving related languages are important:

[4]Hindi is an Indo-Aryan language spoken primarily in North India.
[5]Hausa is an Afro-Asiatic language spoken primarily in Niger and Nigeria.
[6]https://en.wikipedia.org/wiki/Demographics_of_the_European_Union
[7]https://en.wikipedia.org/wiki/South_Asia

1. Translation between related languages *e.g.,* Hindi-Marathi, Malayalam-Bengali (Indic languages), Spanish-French, Bulgarian-Greek (European languages).

2. Translation from a set of related languages to/from an unrelated language. The unrelated language is typically a *lingua franca* for people speaking related languages. *e.g.,* English to/from Indian languages, French to/from Niger-Congo and Afro-Asiatic languages in West Africa.

Many of these languages have few data resources like parallel corpora and/or linguistic resources like morphological analyzers, parsers, *etc.* Since there may be many related languages to support, building resources for each related language/language-pairs may not be feasible. This monograph addresses the problem of translation involving related languages when resources available are limited. *The major thrust of this monograph is the utilization of language relatedness to reduce resource requirements for statistical machine translation.*

1.3 Language Relatedness: Origins and Key Properties

To understand how language relatedness can be utilized, we first summarize the origins of language relatedness and describe key properties that related languages share and are relevant to machine translation.

Throughout this section and the monograph, we have illustrations in Indian languages. We use the BrahmiNet ITRANS romanization scheme (an extension of the ITRANS[8] scheme) to transcribe the examples in Roman script in addition to the native script. Appendix A gives a reference to the transcription scheme we use. When pronunciation is required, we use the *Indian Language Speech sound Label set* (ILSL)[9] (Samudravijaya and Murthy, 2012).

1.3.1 Origins of language relatedness

We describe two major sources of language relatedness in this section.

1.3.1.1 Genetic relatedness

A set of languages is said to be genetically related if they have descended from a common ancestor language. Two languages in the group may have an ancestor-descendant relationship or they may share a common ancestor. The relatedness between languages in the group can be viewed as a tree. Such a group of languages is called a *language family* and the common ancestor of this family tree is called the *proto-language.*

[8] https://en.wikipedia.org/wiki/ITRANS

[9] https://www.iitm.ac.in/donlab/tts/downloads/cls/cls_v2.1.6.pdf

Meaning	Bengali	Assamese
truth	সত্য (satya,/saty/)	হত্য (hatya,/**h**aty/)
assamese	অসমিয়া (asamiyaa,/asamiyaa/)	অহমিয়া (ahamiyaa,/a**h**amiyaa/)
happiness	সুখ (sukha,/sukh/)	হুখ (hukha,/**h**ukh/)
Meaning	**Marathi**	**Hindi**
season	ऋतु (RRitu,/**ru**tu/)	ऋतु (RRitu,/**ri**tu/)
heart	हृदय (hRRidaya,/h**ru**day/)	हृदय (hRRidaya,/h**ri**day/)
sage	ऋषि (hRRiShi,/**ru**sxi/)	ऋषि (hRRiShi,/**ri**sxi/)
Meaning	**Telugu**	**Kannada**
milk	పాలు (paalu,/**p**aalu/)	ಹಾಲು (haalu,/**h**aalu/)
pig	పంది (paMdi,/**p**andi/)	ಹಂದಿ (haMdi,/**h**andi/)
village	పల్లెలు (pall.elu,/**p**allelu/)	ಹಳ್ಳಿಗಳು (haLLigaLu,/**h**alxlxgalxu/)
Meaning	**Hindi**	**Bengali**
government	सरकार (sarakaara,/**sa**rkaar/)	সরকার (sarakaara,/**shax**rkaar/)
sea	सागर (saagara,/**saa**gar/)	সাগর (saagara,/**shaa**gar/)
name	सावित्री (saavitrii,/**saa**vitrii/)	সাবিত্রী (saabitrii,/**shax**bitrii/)

TABLE 1.1: Examples of words showing regularity of sound change among genetically related languages.

The study of genetic relatedness is the subject matter of comparative linguistics. Comparative linguists have studied a large number of the world's languages, both extinct and extant, and have posited a number of language families[10]. Based on historically available records, comparative linguistics have proposed reconstructions of family trees that trace the genetic relationships between languages. These reconstructions are based primarily on the *comparative method*, which uses the principle of regularity of sound change to posit relationships between words (Bynon, 1977). Table 1.1 shows examples of systematic sound changes in some Indian languages. Each word is shown in native script along with (transliteration,/pronunciation/). BrahmiNet-ITRANS is used for transliteration, while ILSL is used for pronunciation. The sound changes are shown highlighted in the pronunciation. For instance, the sound /p/ changes to /h/ going from Telugu to Kannada (both are Dravidian languages).

In this process, they posit *proto-languages*, hypothetical or reconstructed languages from which actual languages are supposed to have descended. For instance, Indo-European is a major language family that encompasses major languages in Europe, Central and South Asia. These languages are believed to have descended from the reconstructed Proto-Indo-European language. Table 1.1 shows examples of some words in a few Indo-European languages which exhibit regularity of sound change.

Are all the languages families descended from a single ancestor language? We cannot answer this question conclusively since the comparative method is dependent

[10]`https://en.wikipedia.org/wiki/List_of_language_families`

Hindi	Gujarati	Marathi	Bengali	Meaning
रोटी (*roTI*)	રોટલો (*roTalo*)	चपाती (*chapAtI*)	রুটি (*ruTi*)	bread
मछली (*maChlI*)	માછલી (*mAChlI*)	मास (*mAsa*)	माछ् (*mACha*)	fish
भाषा (*bhAShA*)	ભાષા (*bhAShA*)	भाषा (*bhAShA*)	ভাষা (*bhAShA*)	language
दस (*dasa*)	દસ (*dasa*)	दहा (*dahA*)	दश (*dasha*)	ten

TABLE 1.2: Examples of cognates in Indo-Aryan languages.

Sanskrit Word	Dravidian Language Word	Language	Meaning
चक्रम् (*cakram*)	சக்கரம் (*cakkaram*)	Tamil	wheel
मत्स्यः (*matsyaH*)	మత్స్యలు (*matsyalu*)	Telugu	fish
अश्वः (*ashvaH*)	ಅಶ್ವ (*ashva*)	Kannada	horse
जलम् (*jalam*)	ജലം (*jala.m*)	Malayalam	water

TABLE 1.3: Examples of some words borrowed from Sanskrit into Dravidian languages.

on historical records to reconstruct family trees, and the available records are not sufficient to trace back the evolutionary tree beyond a certain point in history.

On the other hand, within a language family, we can identify sub-families which correspond to subtrees in the evolutionary tree. These sub-families may show more commonalities since the languages in the sub-trees have diverged from each other at a later point in history. Sometimes, it may be useful to restrict our consideration of language relatedness to a subtree. For instance, the Indo-Aryan, Romance and Slavic are some major sub-families in the Indo-European language family. These sub-families have diverged to a great extent over time. As an example, Indo-Aryan and Romance sub-families exhibit divergence in a property as fundamental as word order: Indo-Aryan languages are SOV (Subject-Object-Verb) languages, whereas Romance languages are SVO languages (Subject-Verb-Object).

As a result of genetic relatedness, related languages share many features. One of the most important features is the presence of *cognates*, which refers to words having a common etymological origin. Table 1.2 shows examples of a few cognates in some Indo-Aryan languages.

1.3.1.2 Contact relatedness

Contact among languages over a long period of time is another major source of language relatedness. Interaction over a long period of time leads to the borrowing of vocabulary (*loanwords*) and adoption of grammatical features from other languages. If such exchange is sustained over a long period of time between languages spoken in contiguous areas, it leads to the formation of *linguistic areas*. In linguistic areas, the languages undergo convergence to a large degree in terms of their structural features (Thomason, 2000). The languages need not belong to the same language family.

Some examples of such linguistic areas are the Indian subcontinent (Emeneau, 1956), the Balkans (Trubetzkoy, 1928) and Standard Average European (Haspelmath, 2001). In the Indian subcontinent, we see a convergence between languages of the Indo-Aryan, Dravidian, Austro-Asiatic (Munda branch) and Tibeto-Burman language families. To cite a few instances of convergence (Subbārāo, 2012):

- Indo-Aryan and Dravidian languages have borrowed vocabulary from each other. Table 1.3 shows some examples of words borrowed into Dravidian languages from Sanskrit, an Indo-Aryan language.

- Indo-Aryan languages have incorporated retroflex sounds and echo words from Dravidian languages.

- The languages of the Munda branch of the Austro-Asiatic family have changed from SVO word order to SOV word order.

In the monograph, we focus on related languages spoken in contiguous areas: (i) genetically related sub-families like Indo-Aryan, Slavic, Indo-Iranian, *etc.* rather than looking at an entire family like Indo-European which is very diverse and (ii) languages which exhibit contact relatedness leading to the formation of linguistic areas like the Indian subcontinent, the Balkans, South-East Asia, *etc.*

1.3.2 Key properties of related languages

In this section, we briefly describe some of the key properties of related languages that are relevant to machine translation.

1.3.2.1 Lexical similarity

Lexical similarity means that the languages share many words with the similar form (spelling/pronunciation) and meaning *e.g., blindness* is represented by the word अन्धापन (*andhapana*) in Hindi and आन्धळेपणा (*aandhaLepaNaa*) in Marathi. Related languages could have different kinds of lexically similar words like:

- **Cognates**: These are words having a common etymological origin *e.g., bread* is रोटी (*roTI*) in Hindi and રોટલી (*roTalI*) in Gujarati.

- **Loan words**: These are words borrowed from one language into another; they are not the result of genetic relatedness of languages. *e.g.,* the word for *wheel* in Tamil is சக்கரம் (*cakkaram*). It has been borrowed from Indo-Aryan languages; in Sanskrit, the word is चक्र (*cakra*).

- **Named Entities**: Named entities are obviously words with similar pronunciation across languages and stand for a particular entity in the real world. *e.g., Kerala* is written as केरल (*kerala*) in Hindi and കേരള (*keraLa*) in Malayalam.

- **Fixed Expressions**: Even fixed expressions like idioms which are culture-specific may be borrowed across languages. *e.g.,* The Hindi idiom दाल में कुछ

काला होना (*dAla me.n kuCha kAlA honA*) and the Gujarati idiom દાળ મા કાઈક કાળુ હોવુ (*dALa mA kAIka kALu hovu*) are very similar lexically and essentially mean the same — *something is suspicious* though it literally translates to *something is black in the lentils*.

There exist words which are similar in pronunciation and spelling but they have different meanings. *e.g.*, The word കാക്ക (*kAkka*) in Malayalam means a *crow*, but the similar sounding word काका (*kAkA*) in Hindi means *paternal uncle*. Such word pairs are called **false friends** and can pose a problem for machine translation.

1.3.2.2 Orthographic similarity

We say that two languages are **orthographically similar** if they have:

- highly overlapping phoneme sets
- mutually compatible orthographic systems
- similar grapheme to phoneme mappings

Sometimes, related languages exhibit orthographic similarity. For instance, Indo-Aryan languages largely share the same set of phonemes. Many Indo-Aryan languages use different Indic scripts derived from the ancient Brahmi script, but correspondences can be established between equivalent characters across scripts. For example, the Hindi (Devanagari script) character क (*ka*) maps to the Bengali ক (*ka*) which stands for the consonant sound (IPA: k). Interested readers may refer the Unicode charts of the two scripts (Devanagari[11] and Bengali[12]) to see the high overlap between the graphemes. The grapheme to phoneme mapping is also consistent for equivalent characters. However, there are exceptions where orthographic similarity may not hold for related languages. For instance, Urdu is written in a script of Arabic origin and the characters do not have a one to one correspondence with Devanagari characters.

1.3.2.3 Structural correspondence

Structural correspondence means that languages have the same basic word order *viz.* SOV (Subject-Object-Verb), SVO (Subject-Verb-Object), *etc.* The basic word order also determines other syntactic properties like the relative order of noun-adposition, noun-relative clause, noun-genitive, verb-auxiliary, *etc.* Related languages typically tend to possess structural correspondence. Divergence in word order is one of the major challenges in machine translation and solutions have been widely investigated. *Because of the nature of the languages involved, word order is not a concern for related languages.*

---

[11] `https://unicode.org/charts/PDF/U0900.pdf`
[12] `https://unicode.org/charts/PDF/U0980.pdf`

Hindi Post-position	Marathi Suffix	Case Description
को (*ko*)	ला (*lA*)	Accusative
को (*ko*)	ला (*lA*)	Dative
से (*se*)	नी (*nI*)	Instrumental
मे (*me*)	त (*ta*)	Locative
का (*kA*)	चा (*cA*)	Genitive

TABLE 1.4: Examples of morphological isomorphism in Marathi and Hindi.

1.3.2.4 Morphological isomorphism

While content words are borrowed or inherited across related languages, function words are generally not lexically similar across languages. However, function words in related languages (whether suffixes or free words) tend to have a one-one correspondence to a large extent. This could be a result of the similarities in the case-marking systems of related languages. Table 1.4 shows some examples of correspondence between suffixes in Marathi and post-positions in Hindi.

1.4 Do We Need SMT Approaches Customized for Related Languages?

Are standard SMT approaches like phrase-based SMT (PBSMT) not sufficient to handle related languages? The promise of statistical machine translation is language independence *i.e.,* the same method should be applicable to any arbitrary pair of languages given parallel corpora for learning these translation systems.

Bender (2011) defines language independence as:

Definition 1.4.1. If technology developed for one language can be ported to another language merely by amassing appropriate training data in the second language, then the effort put into the development of the technology in the first language can be leveraged to more efficiently create technology for other languages.

However, are SMT models really language independent? Let us consider the example of the PBSMT model, a very strong SMT baseline model that is very popular and is often the building block for more sophisticated SMT models. On the face of it, the translation model and distortion model should be able to learn many-many word translation mappings and word reordering phenomena in a language independent manner. However, a closer examination reveals the PBSMT model is more suitable for some language pairs compared to other language pairs. For instance:

- Theoretically, the distortion model can account for the movement of phrases; but it is extremely difficult to model distortion of arbitrary (non-linguistic) phrase pairs. Distance-based and lexicalized distortion models cannot capture complex syntactic divergences among languages. Hence, PBSMT tends to work better between languages with the same word order.

- Though the PBSMT model makes no direct assumptions about morphology, the translation probabilities can be estimated reliably only if enough bilingual phrase pairs are present. This puts morphologically rich languages (synthetic morphology) at a disadvantage since the number of word types is very large and the morphological system is productive enough to generate potentially infinite number of word types. They encounter data sparsity with the result that many words remain untranslated or cannot be reliably translated. Hence, PBSMT tends to work better for languages with isolating or weak inflectional morphology.

To alleviate these limitations, we have to take recourse to language specific processing:

- Modelling word order requires syntactic parsers in order to implement syntax-based SMT (Yamada and Knight, 2001) or source-side pre-ordering for PBSMT (Collins et al., 2005).
- Either we need larger parallel corpora to achieve high vocabulary coverage or we need morphological analyzers to segment the words into constituent morphemes in order to reduce data sparsity.

In that case, SMT cannot be considered language-independent since language-specific resources and tools have to be created if we need to create reasonably good quality translation systems. Thus, between arbitrary language pairs, complete language independence is not possible.

In the extreme case, it may be necessary to develop resources for each language/language-pair in order to cover all related languages in a family. Such an endeavour is clearly not feasible. As we have already discussed, many related languages have limited parallel corpora and other linguistic resources. We hypothesize that these obstacles can be overcome by utilizing the relatedness of the languages involved to effectively use the available resources. Since we know that the languages are related, we can make suitable assumptions in our model to direct the learning process. These serve to compensate for the limited availability of data. In particular, we investigate the following hypotheses in this monograph:

1. **Lexical similarity** can be utilized to overcome the knowledge acquisition bottleneck in low-resource translation between related languages. In the context of translation, knowledge acquisition refers to the ability to acquire bilingual translations from parallel corpora, and this is constrained by the available parallel corpus.
2. Parallel corpora from **multiple related languages** can be pooled together to make effective use of scarce training data. Given the linguistic similarities between related languages, the corpus in one language can be considered to be a

noisy version of an equivalent corpus in another language. Moreover, building translation systems which utilize corpora from multiple related languages may help achieve better generalization.

3. Resources developed for one language can be **ported or re-used** for another related language.

In a nutshell, we explore a **middle-of-the-road solution between (i) building completely language independent translation systems and (ii) building language-pair specific systems**. Instead, we develop translation systems which are partially language-independent *i.e.,* independent within a set of related languages. We consider related languages to be a *suitable grouping of languages* where common solutions can be built and linguistic resources as well as parallel corpora can be shared. While language-independent systems need a lot of data, language-specific systems require a lot of linguistic resources. With related languages, we can make appropriate linguistic assumptions to reduce data requirement as well as create natural language processing tools which are usable across all languages in the set.

1.5 Translation, Transliteration and Related Languages: The Connection

Along with translation, we also address the problem of transliteration between related languages in this monograph. **We believe that translation between related languages and transliteration are complementary problems**:

- Utilizing lexical similarity to improve translation between related languages is one of the major themes of the monograph. Utilizing lexical similarity involves subword-level transformations *i.e.,* the transformation of units more fine-grained than words. Transliteration is also concerned with subword-level transformations, generally character-level.
- In fact, we could consider translation between related languages as a task on a *continuum from translation between arbitrary languages to transliteration.* Translation between arbitrary languages generally involves word/morpheme-transformations, while transliteration can generally be modelled well using character-level transformations. Translation between related languages occupies an intermediate space, where we may need word-level transformation in some cases and character-level transformation (subword-level translation, to be more precise) in other cases.
- Improvements in transliteration systems can lead to improvements in translation between related languages. For instance, a popular method to handle translation of cognates is to transliterate them.

- Transliteration is a relatively simple problem characterized by short sequences and a small vocabulary. It is, therefore, a *good testbed* to experiment with computationally intensive solutions and ideas, before these are explored for translation between related languages.

1.6 What Does the Monograph Contain?

In this monograph, we describe methods to improve translation/transliteration involving related languages by utilizing various similarities between these languages. Specifically, we look at scenarios where resources and tools required for developing translation/transliteration systems are limited (*e.g.,* parallel translation corpora, parallel transliteration corpora, morph-analyzers, *etc.*). *We seek methods which improve translation quality and transliteration accuracy in such low-resource scenarios, enable easy sharing and portability of resources.* The work described in the monograph can be divided into two parts: **Machine Translation** and **Machine Transliteration**. The remainder of this section summarizes the research findings reported in this monograph.

1.6.1 Machine translation

Utilization of lexical similarity by designing appropriate subword-level translation units is a key theme of our work.

1.6.1.1 Subword representation for SMT

Lexical similarity is a key property of related languages, which establishes subword-level correspondences between translations. We can utilize lexical similarity to reduce data requirements by exploring subword-level transformation *e.g.,* पानी (pAnI) and पाणी (pANI) are cognates for *water* in Hindi and Marathi, respectively, and translation only involves transformation of a single character.

We explore *phrase-based SMT with subword units* (*i.e.,* translation units more fine-grained than the word) to utilize lexical similarity. We analyze the limitations of subword units proposed in previous work and show improvements due to the use of two translation units which had previously not been explored for SMT between related languages: Orthographic syllables (OS) and Byte Pair Encoded units (BPE units).

1. **Orthographic Syllable**: We propose the syllable as a translation unit between related languages. Syllables are phonological *building blocks* of words which can capture phonotactic constraints of a language. True syllabification is non-trivial due to problems like coda-onset ambiguity. For instance, the word *abhijit* can be syllabified as *a-bhi-jit* or *ab-hi-jit*. Hence, we propose

to use an approximate syllable, which we refer to as the *orthographic syllable*, as the translation unit. Orthographic syllabification can be performed deterministically.

2. **Byte Pair Encoded Unit**: Syllables represent one kind of subword-level pattern for which we want to learn bilingual mappings. However, other kinds of patterns are of interest too: (i) linguistic patterns like morphemes, suffixes, prefixes, (ii) frequent non-linguistic patterns. Such interesting patterns are characterized by their frequency of occurrence. Hence, we propose the use of such frequent subwords as translation units. We use Byte Pair Encoding (Gage, 1994), a text compression algorithm, to identify such frequent subwords. This algorithm provides a well-motivated statistical procedure to identify subwords.

We show that OS and BPE units significantly outperform previously proposed translation units *viz.* characters, character n-grams, morphemes and words. Through quantitative and qualitative analysis, we show that the improvement in translation quality due to OS and BPE units can be attributed to the following reasons: (i) reduction in data sparsity due to smaller and finite vocabulary, (ii) judicious balance in use of lexical similarity and word-level information and (iii) ability to learn diverse lexical units.

1.6.1.2 Pivot Translation with Subword Representation

We explore another resource-scarce scenario: SMT between two related languages that do not share parallel corpus, but they share parallel corpora with a common third language, referred to as *pivot language*. Translation between the two languages is mediated by the pivot language, a paradigm referred to as *pivot-based SMT* (Utiyama and Isahara, 2007). We propose a pivot-based SMT model which can utilize lexical similarity between the source, pivot and target languages to improve translation quality. This is achieved by: (i) using a pivot language related to the source and target languages and (ii) subword-level translation units discussed earlier (OS and BPE units). More significantly, we show subword-level pivot translation models are competitive with direct translation models[13].

Beyond pivot-based models, we investigate multilingual translation models built using pivot-based SMT *i.e.,* translation between two languages assisted by multiple pivot languages. We show that combining multiple subword-level pivot models, one for each pivot language, can lead to a translation model combination whose translation quality is nearly equivalent to that of the best direct translation model.

1.6.1.3 Reusing resources across related languages

We show that resources created for one language can be reused for other related languages. For this study, we consider source-side pre-ordering as a case study. Source side pre-ordering (Collins et al., 2005) is a well-known method for addressing

[13]Direct translation refers to the translation model learnt using a parallel corpus between the source and target languages.

word-order divergence between source and target language while training PBSMT models. In this approach, the source-side training corpus is preprocessed such that the word order resembles the word order of the target language. We show that the same rule-base can be used successfully for multiple Indian languages since Indian languages exhibit structural correspondence.

1.6.1.4 MT and contact relationship

Previous work on SMT involving related languages has been limited to languages that share a genetic relationship (Nakov and Tiedemann, 2012; Durrani et al., 2010). In this monograph, we expand our study to include languages that share a contact relation but are not necessarily genetically related. We show that all the approaches to utilizing linguistic similarities that we propose in the monograph can also benefit languages that are related by contact.

1.6.1.5 Case study on indic language translation

The relatedness between Indian languages is a major motivation for the work presented in this monograph. Hence, we conduct an extensive case-study on translation involving nine major Indian languages, covering 72 language pairs and spanning two language families. We experiment with the different translation units discussed previously. We also report English to Indian language as well as Indian language to English translation results. The case-study thus spans 90 language pairs (72 Indian-Indic, nine English-Indic and nine Indic-English language pairs).

1.6.2 Machine transliteration

In the case of transliteration, we show how **orthographic similarity** can be utilized to improve transliteration between related languages. We focussed on two challenging scenarios where orthographic similarity could be beneficial: (i) unsupervised transliteration and (ii) multilingual transliteration. We look at the proposed solutions as stepping stones towards unsupervised translation and multilingual translation for related languages.

1.6.2.1 Substring-based unsupervised transliteration

In the unsupervised transliteration task, we learn a transliteration model between two related languages without a parallel transliteration corpora, relying only on monolingual words list in the two languages. We proposed a two-stage, iterative bootstrapping approach to utilize the orthographic similarity between languages for improving unsupervised transliteration.

1.6.2.2 Multilingual transliteration

In the multilingual transliteration task, we jointly train multiple transliteration pairs involving related languages. We can interpret this as a multi-task problem, where every transliteration pair corresponds to a task. The orthographic similarity between related languages can be interpreted as task similarity and helps improve transliteration quality.

We propose a compact neural encoder-decoder architecture, that is designed to ensure maximum sharing of parameters across languages while providing room for learning language-specific parameters. This allows greater sharing of knowledge across language pairs by utilizing orthographic similarity. This architecture helps indirectly share limited parallel transliteration corpora among multiple related language pairs. It also leads to better generalization since the network prefers to learn representations that work well across multiple languages.

Chapter 2

Past Work on MT for Related Languages

In this chapter, we present a survey of the past work regarding machine translation involving related languages. In the previous chapter, we saw that there are two major use-cases for translation involving related languages. The major part of the survey is organized along these use-cases. Section 2.1 looks at the literature related to translation between related languages. Section 2.2 looks at the literature related to translation between a set of related languages on one hand and a *lingua franca* (link language) on the other hand. These sections are mostly focussed on statistical machine translation, which has been the pre-eminent machine translation paradigm in the last couple of decades (and the focus of most of the work in this monograph). In addition, we summarize the nascent work and potential future directions for translation involving related languages in the emerging area of Neural Machine Translation (Section 2.3). We also look back at the work on MT for related languages in the context of rule-based systems (Section 2.4).

2.1 Translation between Related Languages

This section covers relevant work on translation between two related languages which utilizes lexical similarity in order to improve statistical machine translation.

2.1.1 Corpus augmentation with lexically similar word pairs

Addition of lexically similar word pairs (cognates, loan words, etc.) to the available corpus is the simplest way to utilize lexical similarity. Lexically similar word pairs can be mined using a variety of metrics to quantify lexical similarity like character-level Normalized Edit Distance, Longest Common Subsequence Ratio, *etc.* (Inkpen et al., 2005; Kondrak, 2000; Covington, 1996; Melamed, 1995; Wagner and Fischer, 1974). Addition of lexically similar word pairs increases the vocabulary coverage of the translation system. Kondrak et al. (2003) show that this simple method also improves word alignment. As a result, Kondrak et al. (2003) observe modest improvement in overall translation quality. A few simple tricks can provide some improvements in word alignment. A high recall system to extract lexically similar words has been shown to a more beneficial than a high precision system (Kondrak et al., 2003; Al-Onaizan et al., 1999). Instead of including the lexically similar word pairs only once in the augmented parallel corpus, they can be replicated multiple times (Kondrak et al., 2003). Having multiple word pairs in a sentence forming a pseudo-parallel sentence is marginally better than have one sentence for every word pair (Kondrak et al., 2003).

This method can only translate lexically similar words which are part of the augmented corpora, hence the gains seen are limited by the number of lexically similar word pairs added to the original corpus. It cannot handle translation of out-of-vocabulary words encountered at test time.

2.1.2 Using lexical similarity features for word alignment

The popular generative word alignment models like the IBM Models (Brown et al., 1993) learn word translation probabilities based on co-occurrence of word pairs in the parallel corpus. However, there are other important signals which may help improve word alignment *e.g.,* relative positions of words in the source and target sentence, part-of-speech tags, *etc.* Discriminative alignment models (Moore, 2005) allow the incorporation of such arbitrary features. Taskar et al. (2005) have used binary features capturing lexical similarity like: do source and target words (a) match exactly, (b) match ignoring accents, (c) match ignoring vowels?

They also use lexical similarity metrics like LCSR (Melamed, 1995) as features to incorporate lexical similarity. As we have seen, short words can be problematic. So they incorporate a binary feature to distinguish long and short words. They report a 7% reduction in the alignment error rate for English-French using these features. However, this method too cannot handle translation of out-of-vocabulary words.

2.1.3 Transliteration of lexically similar words

Transliterating untranslated words in the output of an SMT system (a post-processing step) has been a common strategy to handle unknown named entities in statistical machine translation (Al-Onaizan and Knight, 2002; Hermjakob et al., 2008; Kashani et al., 2007). Lexically similar words can be handled in a similar way. Since lexically similar words are spelt similarly, we can hope that transliteration of untranslated words would generate the correct translations. Nakov and Tiedemann (2012) and Durrani et al. (2010) show that transliteration of untranslated words can improve translation quality. This method can handle OOV words since all untranslated words are transliterated.

Translation choices depend on the contextual words in candidate output sentence. The proposed post-processing method does not allow the candidate translations generated by the transliteration system to be evaluated in the context of other words in the candidate output sentence. This can be addressed by considering the top-k transliteration candidates and then re-ranking each possible candidate output sentence (generated by substituting the OOV with transliterations) using a target language model (Durrani et al., 2014).

Yet, the transliterations cannot be scored and tuned along with other features used in the SMT system. One way of partially mitigating this limitation is a two-pass decoding step, where transliterations of untranslated words are plugged back into the source and a second decoding run is performed.

Integration of the transliteration module into the SMT decoder is a more elegant approach to overcome the above-mentioned limitations (Durrani et al., 2010, 2014). Transliteration is performed on the fly, and both translation and transliteration candidates can be evaluated and scored simultaneously. This also allows transliteration *vs.,* translation choices to be made.

Other than the above-mentioned works, the use of transliteration for translation between related languages is largely unexplored — with most work focussing on handling named entities.

This method has a couple of limitations. First, a transliteration model trained on named entities (as is generally done) cannot capture the subword-level transformations involving lexically similar words. This issue can be overcome by mining parallel transliteration corpora from the parallel translation corpora (Sajjad et al., 2012). The mined transliteration corpus is more representative of the subword-level transformations between lexically similar words. Second, a character-level transliteration model is not powerful enough to model the translation of lexically similar words.

2.1.4 Character-level SMT

The approaches discussed in the previous section essentially retrofit a word-level SMT model with tweaks to incorporate lexical similarity. A radically different translation paradigm involves building translation systems that use more granular subword-level translation units. Characters, character n-grams have been explored to leverage lexical similarity and reduce data sparsity. Character-level SMT has been explored for very closely related languages like *Bulgarian-Macedonian, Indonesian-Malay, Spanish-Catalan* with modest success (Vilar et al., 2007; Tiedemann, 2009; Tiedemann and Nakov, 2013). These results were demonstrated primarily for very close European languages.

While building a character-level SMT system, the standard SMT training procedure needs a few adaptations:

- A higher-order language model is used compared to word-level LMs to incorporate larger context. Since the size of the vocabulary (set of characters) is small, data sparsity is not a problem Tiedemann (2012a).
- During tuning, it is better to optimize the word-level BLEU objective rather than character-level BLEU objective (Nakov and Tiedemann, 2012).

The primary drawback of character-level SMT is the limited contextual information unigrams provide for learning translation models (Tiedemann, 2012a). Hence, some work has explored higher-order character n-grams. However, the use of character n-gram units to address this limitation leads to data sparsity for higher-order n-grams and provides little benefit beyond (n = 2) (Tiedemann and Nakov, 2013).

We would also like to point out that another subword-level unit *viz.* the morpheme has typically been used as a translation unit to mitigate data sparsity when the languages involved are morphologically rich (Lee, 2004). Both hand-crafted/supervised morphanalyzers (Lee, 2004; Goldwater and McClosky, 2005) as well as unsupervised morphanalyzers (Virpioja et al., 2007) have been employed for morpheme segmentation for SMT. Unsupervised morphanalyzers are particularly appealing for low-resource languages since they can learn segmentations from monolingual corpora alone and require no additional resources. Morpheme-level translation is a general method and it does not require the source and target to be lexically related. Hence, it does not utilize lexical similarity. Most work on using morpheme-level translation has not studied its utility for related languages. We believe that morphological isomorphism between related languages can make morpheme segmentation particularly useful for related languages.

2.2 Translation Involving Related Languages and a Lingua Franca

This section summarizes the work related to translation involving a set of related languages on one hand, and an unrelated language (possibly a *lingua franca*) on the other hand. In the simplest setting, we can concretely characterize this problem as: We would like to translate from a resource-poor language X to an unrelated language E but X and E share no parallel corpus. But, language Y (which is related to language X) is resource-rich and shares a parallel corpus with language E. This use-case results in the different scenarios, depending on the availability of parallel corpora, which are listed below:

- The related languages X and Y share a small amount of any parallel corpus
- The related languages X and Y do not share any parallel corpus

The solutions proposed for these scenarios in previous work rely on using Y as the pivot language to translate from X and E. The reverse problem of translating from E to X via Y can also use the same solutions. In the remainder of this section, we summarize the proposed solutions to these problems.

2.2.1 Related languages share some parallel corpus

Suppose related languages X and Y share some parallel corpus. To address this scenario, building a pivot-based SMT system from X to E using Y as the pivot language is the obvious solution. Since we have a limited parallel corpus between X and Y, we could utilize any of the techniques described in Section 2.1 to build a machine translation system between X and Y.

Most of the reported work has built character-level SMT models between X and Y to utilize lexical similarity, and word-level SMT models between Y and E. These works use different approaches to build a pivot-based SMT system from X to E (referred to as the X $\xrightarrow{char}$ Y $\xrightarrow{word}$ E system): (i) Pipelining, (Tiedemann, 2012a; Tiedemann and Nakov, 2013) and (ii) Synthetic corpus generation (Tiedemann and Nakov, 2013). Note that triangulation cannot be used to build the X-E pivot-based SMT system since the X-Y and Y-E systems are trained with different translation units.

These works have reported results with Slavic, Iberian and Nordic languages as related language groups and English as the unrelated language. The following is a summary of the major observations:

- They show that this approach beats a baseline system which uses a word-level X-Y translation system (referred to as the X $\xrightarrow{word}$ Y $\xrightarrow{word}$ E system).
- Tiedemann (2012a) shows improvement in the E-Y-X direction as compared to the X-Y-E. The performance of X-Y-E may degrade since the X-Y character-level system may generate non-words, which the Y-E system cannot translate. However, the authors did not find too many non-words and suggested further investigation.
- Tiedemann (2012a) also shows the utility of this model for domain-adaptation for Nordic languages. If domain-specific corpus for X-E is not available and Y-E domain-specific corpus is available, they show that the proposed X-Y-E system can outperform a translation system trained on an out-of-domain X-E parallel corpus.
- Tiedemann and Nakov (2013) observe that synthetic corpus generation performs better compared to pipelining.

2.2.2 Related languages share no parallel corpus

Suppose related languages X and Y share no parallel corpus. To address this scenario, Wang et al. (2012, 2016) explore *source adaptation i.e.,* rewriting the Y part of the Y-E corpus to language X (say corpus X'), and then training the newly created X'-E corpus. This is similar to synthetic corpus generation for pivot-based SMT, but the X' corpus has to be generated without access to an X-Y parallel corpus. Instead, other resources are used to utilize lexical similarity and to generate translation candidates for words and phrases in Y:

- Bilingual dictionaries
- Transliteration of words in Y to X
- Morphological variants of Y for language X are generated using morphological analyzers for Y and morphological generators for X
- If a small X-E corpus is available, X-Y word and phrase translations are induced from the X-E and Y-E by pivoting around E

Given a sentence in Y, translation candidates are generated for every word in X along with scores. This forms a confusion network[1] (a simpler variant of a lattice) representing a large search space of possible translations. In addition to translation of each word, we also have to consider the fluency of the output sentence which can be scored using a language model for X. Thus, the best translation can be found by finding the best path through the confusion network which maximizes a function that combines the translation and language model scores. Wang et al. (2012) propose a simple decoder which uses a language model of the language X to find the best adaptation of Y to X in the confusion network. Wang et al. (2016) propose an improved, iterative rewriting decoder which can incorporate more features in addition to language model scores to find the best adaptation. These features include transliteration model scores, translation probabilities for words/phrases (if available) and sentence-global features like length, type-to-token ratio *etc.*

The following are the major observations:

- They show that the proposed source adaptation method is better than a baseline system where the sentences in X are input to the Y-E system *i.e.,* in the absence of an X-Y corpus, the Y-E system pretends to be the X-E system.

- They also show that *reverse adaptation* is not as effective as source adaptation. Reverse adaptation refers to the adaptation of a sentence in X to a sentence in Y at decode time, before using the Y-E translation system (essentially a pipelined pivot translation system). It seems source adaptation method can better respond to noisy translations.

- Wang et al. (2016) also show that the feature-rich rewriting decoder is better than the language-model based decoder.

2.2.3 Limited parallel corpus between unrelated languages

So far, we have looked at cases where no X-E parallel corpus is available. If a small X-E is also available, then the source adapted model (X'-E) and the model (X-E) trained on the small parallel corpus can be combined/ensembled to create a better model. Many methods have been proposed in the SMT literature to combine two translation systems: (a) combination of phrase-tables along with linear interpolation of feature values (Wu and Wang, 2009), (b) concatenation of phrase-tables with extra features to identify origin of each phrase pair (Nakov and Ng, 2009), (c) fill-up interpolation (a backoff scheme which defines the priority of phrase-tables being combined) (Dabre et al., 2015), (d) decoding time combination which searches over multiple paths (Nakov and Ng, 2009; Dabre et al., 2015). These methods are not specific to related languages.

Since X and X' are the same language (X' is created through source adaptation), Nakov and Ng (2009) create a new corpus by concatenating X-E and X'-E corpora and then training the combined X-E translation model. All these combination methods yield some benefit over the original model (X-E).

[1] `http://www.statmt.org/moses/?n=Moses.ConfusionNetworks`

However, there is scope for improvement since the combined model may be biased towards the source-adapted model (X'-E) as the X'-E corpus is much larger than the X-E corpus. To counteract this limitation, Nakov and Ng (2009) explore the following strategies to improve the corpus concatenation method:

1. Simple Concatenation: Concatenate one copy of the smaller X-E with one copy of the X'-E corpus.
2. Balanced Concatenation: Concatenate k copies of the smaller X-E with one copy of the X'-E corpus.
3. Balanced Concatenation with Selection: This method tries to improve the word alignment in X-E using X'-E rather than combining the two. They perform balanced concatenation, perform word alignment on the concatenated parallel corpora, then retain only one copy of the X-E word-aligned corpus for further training of the model. The rationale here is that word alignment using the Balanced Concatenation corpus comprising the larger X'-E corpus improves word alignment for the smaller X-E corpus.
4. Sophisticated Concatenation: Phrase tables learnt from models (1) and (3) described above are combined. The combined model benefits from the better X-E alignments in Model (3) and better lexical coverage of Model (1).

Nakov and Ng (2009) report the following major observations:

- The concatenation-based approaches, which utilize lexical similarity, are better than the generic combination methods (a) to (d) mentioned above.
- Balanced concatenation is better than simple concatenation.
- Sophisticated concatenation outperforms the other methods.

2.3 Neural Machine Translation and Related Languages

Neural Machine Translation (Sutskever et al., 2014; Bahdanau et al., 2015) has become an important machine translation paradigm and shown significant improvements in translation quality over SMT for many language pairs. In this section, we summarize some of the recent work on NMT that is relevant to related languages.

2.3.1 Subword-level translation

NMT has seen research on subword-level translation. It has been motivated by a major limitation of NMT systems: a large vocabulary makes *softmax* computation inefficient. Early solutions relied on a truncated vocabulary of the most frequent words, but this meant low-frequency words could not be translated. To overcome this limitation, various researchers have proposed representing words using a smaller translation vocabulary. For instance, the set of characters can be chosen as the translation vocabulary (Chung et al., 2016; Lee et al., 2017). The smaller vocabulary makes softmax computation efficient, and all words in the corpus can be represented as a function of the chosen vocabulary. As a positive side-effect, open-vocabulary translation is possible since unseen words can be generated using the translation vocabulary.

In addition to characters, researchers have proposed other translation units like Byte Pair Encoded units and wordpieces:

- Byte Pair Encoded units (Sennrich et al., 2016): The vocabulary consists of the most frequent character sequences in the training corpus, chosen such that maximum compression of the corpus is achieved. The vocabulary is learnt using the Byte Pair Encoding text compression algorithm (Gage, 1994).
- Wordpieces (Schuster and Nakajima, 2012; Wu et al., 2016): Wordpieces are also character sequences and the wordpiece vocabulary is learnt by finding segmentations of words in the corpus that maximize the likelihood of monolingual corpus.

These translation units are widely used for NMT between arbitrary languages. The fact that these translation units are at the subword-level raises the possibility of using them to utilize lexical similarity for translation involving related languages.

2.3.2 Multilingual translation

In the case of SMT, pivot-based SMT is a standard method to build multilingual translation models. The major limitations of pivot-based approaches are:

1. The models are not trained end-to-end to maximize the source-target translation quality. Rather, source-pivot and pivot-target translation models are trained separately and combined using various heuristics to build source-target translation models.

2. Pivot-based SMT faces data sparsity due to vocabulary mismatch between the pivot language sides of the source-pivot and pivot-target parallel corpora. This degrades the quality of the pivot-based models built from the source-pivot and pivot-target models. For instance, the pipelining method will not be able to further translate a pivot language word in the output of the source-pivot translation step, if the word is not represented in the pivot-target translation model.

3. Character-level pivot models can generate non-words, which can be difficult to process in subsequent stages of the pivot-based MT as discussed in Section 2.2.

Multilingual models using neural networks allow end-to-end training. The distributed representation of words/subwords can overcome problems with data sparsity and generation of non-words. Multilingual neural translation models have been explored in two scenarios: (a) multilingual learning and (b) transfer learning.

2.3.2.1 Multilingual learning

In the multilingual learning scenario, the goal is to learn a single translation model given data from multiple language pairs *i.e.,* joint training. It is an application of multi-task learning, where every language pair is a task. The more related the tasks, the more the sharing of information between the tasks is beneficial. Multilingual training, based on neural network models, hopes to improve translation quality by: (a) pooling together training data from multiple language pairs and (b) learning representations that generalize to many languages.

Different network architectures have been proposed for multilingual learning. These architectures try to maintain a balance between sharing of information across languages and preserving language-specific information in the models. In neural network models, information is captured by the model parameters (embeddings, encoder parameters, decoder parameters, output layer parameters, *etc.*). Various architectures differ in the extent to which different model components like the embeddings, encoder, decoder, *etc.* are shared between languages. Some architectures have minimal sharing (Firat et al. (2016a) share only the attention layers). Zoph and Knight (2016) and Lee et al. (2017) share the encoder across multiple languages, while Dong et al. (2015) share the decoder across multiple languages. At the other extreme, Johnson et al. (2017) share the entire network across all languages. They achieve differentiation among target languages by incorporating a special token in the input stream that indicates the target language. *e.g.,* An input sentence for English-Hindi translation, *I am going home*, will be prefixed by a special token *@HINDI@* to indicate the target language. This meant that the vocabulary contains special tokens for the target languages. Different network architectures have been proposed for multilingual learning. These architectures try to maintain a balance between sharing of information across languages and preserving language-specific information in the models. In neural network models, information is captured by the model parameters (embeddings, encoder parameters, decoder parameters, output layer parameters, *etc.*). Various architectures differ in the extent to which different model components like the embeddings, encoder, decoder, *etc.* are shared between languages. Some architectures have minimal sharing (Firat et al. (2016a) share only the attention layers). Zoph and Knight (2016) and Lee et al. (2017) share the encoder across multiple languages, while Dong et al. (2015) share the decoder across multiple languages. At the other extreme, Johnson et al. (2017) share the entire network across all languages. They achieve differentiation among target languages by incorporating a special token in the input stream that indicates the target language. *e.g.,* An input sentence for English-Hindi translation, *I am going home,* will be prefixed by a special token *@HINDI@* to indicate the target language. This meant that the vocabulary contains special tokens for the target languages.

Sharing of networks parameters can be useful for related languages. Experiments in the reported literature have generally focussed on a set of randomly chosen languages, but it would not be surprising if these methods work very well when parameters are shared between related languages.

2.3.2.2 Transfer learning

In this scenario, knowledge learnt for translation of a language pair is used to improve another translation of another language pair. We train a translation model for a high-resource language pair, and then use the model for a low-resource language pair. *e.g.,* A translation model is trained for the high-resource French-English language pair (*parent model*), and the same model is used to translate from Hausa to English, which is a low-resource language pair. The French language-specific parameters in the model are shared with Hausa. In this case, we have achieved *zeroshot translation, i.e.,* translation between Hausa and English without any parallel corpus. Moreover, if we have a small Hausa-English parallel corpus, then the French-English model can be fine-tuned to perform Hausa-English translation (*child model*). A number of recent papers have proposed this approach of parameter sharing and fine-tuning recently (Zoph et al., 2016; Firat et al., 2016b; Johnson et al., 2017).

Dabre et al. (2017) and Nguyen and Chiang (2017) have explored this idea for related-languages, where a high-resource language is used to improve translation involving a low-resource related language. In order to utilize the lexical similarity between these languages, they also use BPE units as translation units instead of words. Multiple works have shown that transfer learning works best for related languages (Maimaiti et al., 2019; Dabre et al., 2017, 2018). Related high-resource languages can also be useful in adapting a massively, multilingual model to a low-resource scenario — a process referred to as "similar language regularization" (Neubig and Hu, 2018).

2.4 Rule-based MT Systems Involving Related Languages

The previous sections dealt with data-oriented machine translation methods involving related languages. In this section, we summarize the prior work on rule-based machine translation systems involving related languages. Rule-based methods for related languages have been proposed for many language groups. Most of these methods have commonalities with respect to handling language relatedness. The related literature for Indian languages is particularly rich, and spans the entire gamut of paradigms. Hence, we ground this discussion with reference to rule-based MT systems for Indian languages and provide appropriate pointers for literature related to other language groups.

Three rule-based MT systems for Indian languages serve as prototypical examples of methods for handling related languages:

- AnglaBharathi: English to Indian language translation system
- Anusaaraka: Indian to Indian language "accessor"
- Sampark: Indian to Indian language translation system

2.4.1 AnglaBharati

AnglaBharati (*Angla* $\rightarrow$ *English, Bharat* $\rightarrow$ *India*) is an English-Indian language translation system (Sinha et al., 1995), which tries to build a common framework for translation between English and Indian languages. In this method, the analysis of the English text is driven by context-free like parsing. The analysis stage is independent of any target Indian language. This is achieved by transforming the source text into an intermediate pseudo-target representation, which is applicable to the group of Indian languages.

This method is neither a purely transfer-based MT system between two languages, nor a language-independent interlingua-based method. The method and representation are at an intermediate level that achieves independence within the language group, while circumventing the difficult problem of generating a true language-independent representation which resolves all linguistic ambiguities.

The intermediate representation consists of word-groups (which can be thought of as noun, verb, preposition phrases *etc.*). Syntax planning rules ensure that the sequence of the word groups match the target language's word-order. Thus, the common analysis stage can address the structural divergence between English and Indian languages via a common rulebase.

Language-specific text generators convert the pseudo-language text to the target language text. This involves satisfying various constraints like case-marking, agreement, tense, *etc.* Indian languages share a rich tradition of Paninian Grammar. Paninian grammar provides a convenient, common framework to design text generators for Indian languages (Bharati et al., 1996).

The architecture also contemplates a further level of specialization, where different language families in India (Indo-Aryan, Dravidian, Austro-Asiatic and Sino-Tibetan) may have their own pseudo-language representation.

2.4.2 Anusaaraka

Anusaaraka (meaning *to follow*) is an Indian-Indian language "accessing software", rather than a true machine translation system (Bharati et al., 2003). Given that Indian languages are similar in many respects, the system employs simple linguistic processing to transform the source language to a representation that is close to the target language. However, the output in the target may not be fluent in terms of canonical usage, case markings, agreement *etc.* Given the similarity between the two languages, the system pushes some of the burden of cross-lingual communication to the user and expects the user to understand the intended meaning. We could consider this output to be in a *dialect* of the target language, and the user will need some training/acclimatization before he can feel comfortable with the system output. The following example of Malayalam-Hindi translation from the original paper illustrates the idea:

mal: `rAm puRattekk poyI .enn lakShmaN paRaJNJNu`
hin: `rAm bAhara gayA aisA lakShmaN kahA`

gloss: `Ram out went that Lakshman said`
eng: `Lakshman said that Ram went out`

In the above sentence, an intermediate translation is provided, which is almost word-for-word. But some case markers have not been correctly generated. Though the sentence is grammatically correct and understandable to a Hindi speaker, this construction is not typical in Hindi. In Hindi, the sentence would be structured as:

`lakShmaNa ne kahA kI rAma bAhara gayA hai`

The conceptualization of such an *accessor* simplifies its design. Analysis typically consists of morphological analysis, local word grouping and mapping using bilingual dictionaries. Given that fluent target language output is not mandated, the word generator consists of a simple set of rules. Note that no rules for structural transformations are required since the languages share roughly the same word order. The language-specific analyzers and generators are fairly simple to develop. For supporting translation between n Indian languages ($n(n-1)$ translation systems), only n analyzers and n generators need to be developed (one per language).

As the name suggests, *Anusaaraka* aims to provide the user with results of the analysis at every stage, so that the user is presumably able *to follow* the analysis in order to understand the output in case of ambiguity or mistakes.

2.4.3 Sampark

Sampark (meaning *contact*) is an Indian-to-Indian language transfer-based machine translation system (Anthes, 2010). It can be seen as a natural progression from the *Anusaaraka* accessor system to a full-fledged translation system. The overall high-level methodology remains the same but the analysis and synthesis stages are much more involved to ensure correct output generation in the target language. The analysis and synthesis stages are connected by a transfer stage, which mainly performs lexical transfer along with transliteration of unknown words and minimal syntactic transfer. Paninian Grammar provides a common framework across Indian languages for analysis and synthesis.

Similar architectures have also been proposed for Slavic languages (Hajič et al., 2000; Homola and Kubon, 2004; Homola, 2008) (the *Česílko* system) and Turkic languages (CICEKLI, 2002). Homola (2008) apply a hybrid architecture using a language model to re-rank multiple outputs from the rule-based system.

2.4.4 Comparison between rule-based and data-oriented systems

Rule-based and data-oriented systems for related languages are similar in some respects. Both MT paradigms utilize the structural correspondence to simplify transformation rules. Rule-based systems like AnglaBharati share structural transformation rules among multiple languages for English-Indian language translation; the possibility of sharing these rules for English-Indian language SMT is a direction worth exploring. Both MT paradigms also utilize morphological isomorphism: (a) the use of morpheme-level translation reduce data sparsity, (b) morphological isomorphism simplifies lexical transfer and word generation.

The literature on related languages in the two paradigms differs in terms of how language relatedness is primarily utilized.

Rule-based systems are concerned with modularizing the software components so that related languages can share different components (like structure transformation rules) or have a common framework across related languages for developing different components like morphanalyzer (like the Paninian Grammar framework for Indian languages). In short, rule-based systems try to achieve maximal sharing of software components; which is not surprising since development of transfer rules *etc.* constitutes a major chunk of the development of rule-based systems. However, linguistic resources like lexicons, paradigm tables *etc.* still have to be developed for every language.

The literature on data-oriented methods, on the other hand, is focussed on sharing parallel corpora across languages and reducing parallel corpora requirements. To this end, the utilization of lexical similarity is a key component of the data-oriented systems for related languages. Rule-based systems have not utilized lexical similarity, relying instead on bilingual dictionaries for lexical transfer. There is no reason why rule-based systems cannot utilize lexical similarity. However, of the systems we have included in this study, only *Sampark* incorporates a transliteration module to utilize lexical similarity.

2.5 Summary

In this chapter, we presented an overview of prior work on machine translation involving related languages. We covered the literature on SMT, NMT and rule-based MT systems. Following are the major highlights:

1. Utilization of lexical similarity is a major theme in the related work on SMT. Most work has involved character-level transformations, which is insufficient to model lexical differences between related languages.

2. Previous work has generally looked at SMT between very close, genetically related languages. There exists limited work on SMT between languages that are farther apart (like Marathi and Hindi; related but not mutually intelligible).

3. Translation between languages related by contact has not received attention in the SMT literature, but has been addressed in the work on rule-based systems.

4. Pivot-based SMT techniques have been used to assist translation involving a low-resource language using a resource-rich related language.

Part I

Machine Translation

Chapter 3

Utilizing Lexical Similarity by Using Subword Translation Units

In Chapter 1, we identified lexical similarity as a key property of related languages. In this chapter, we show that lexical similarity can be successfully leveraged to improve translation quality between related languages, especially when limited parallel corpus is available. To this end, we propose that appropriate subword representation of bilingual text can help utilize lexical similarity to improve translation quality. By subword representation, we mean that the basic units of translation are sub-components of a word.

Over the next three chapters, we investigate various questions related to subword-level SMT. This chapter and the next one are concerned with subword-level SMT between two related languages when they share a parallel corpus. Chapter 5 explores subword-level translation when the related languages do not share a parallel corpus, but pivot languages are involved.

3.1 Motivation

Why do we need subword-level translation? SMT depends on the ability to learn word-level correspondences by using co-occurrence information in parallel corpora. However, we need a large corpus to reliably estimate these probabilities and achieve good vocabulary coverage. A greater challenge is for morphologically rich languages which are very productive in the number of word forms they can create. Morphological analyzers are not available for many languages. Unsupervised morph analyzers can be a substitute, but they may also result in a large vocabulary size. Yet, we will not be able to do open vocabulary translation. Moreover, morpheme representation can utilize morphological isomorphism but cannot utilize lexical similarity between source and target languages.

Is there a good rationale for using translation units that can utilize lexical similarity? Traditionally, machine translation has relied on learning mappings between meaning bearing linguistic units like words and morphemes. This choice is probably motivated by a fundamental principle of linguistics: arbitrariness between a word's form (orthography and sound) and its meaning (Saussure, 1916). However, does it naturally follow from this principle that the mapping between the forms of similar meaning words across languages is also arbitrary? It may be true for two randomly chosen languages, and hence in the most general setting word and morpheme-level translation is a reasonable design choice. However, in the case of related languages, there is a relationship between the forms of the words – lexical similarity. Since many words are cognates, lateral borrowing and loan words, certain phonological processes connect the forms of similar meaning words in related languages. This relation is also reflected in the orthographic representation of these languages. It is, therefore, reasonable to assume that a word (more precisely its orthographic representation) can be factorized into component subword representations that better reflect the relationship between the word forms across languages. Such a subword representation should be able to learn translation mappings at the subword-level. We believe that such a subword-level representation that reflects the lexical similarity between source and target languages is the right level of abstraction for translation between related languages.

3.2 Related Work

There are two broad sets of approaches that have been explored in the literature for translation between related languages that leverage lexical similarity between source and target languages.

The first approach involves **transliteration of source words** into the target languages. This can be done by transliterating the untranslated words in a post-processing step (Nakov and Tiedemann, 2012), a technique generally used for handling named entities in SMT. However, transliteration candidates cannot be scored and tuned along with other features used in the SMT system. This limitation can be overcome by integrating the transliteration module into the decoder (Durrani et al., 2010), so both translation and transliteration candidates can be evaluated and scored simultaneously. This also allows transliteration *vs.*, translation choices to be made.

Since a high degree of similarity exists at the subword-level between related languages, the second approach looks at **translation with subword level basic units**. Character-level SMT has been explored for *very closely* related languages like *Bulgarian-Macedonian, Indonesian-Malay, Spanish-Catalan* with modest success (Vilar et al., 2007; Tiedemann, 2009; Tiedemann and Nakov, 2013). Unigram-level learning provides very little context for learning translation models (Tiedemann, 2012a). The use of character n-gram units to address this limitation leads to data sparsity for higher-order n-grams and provides little benefit (Tiedemann and Nakov, 2013).

The choice of characters and character n-grams as translation units is *ad hoc*. In this chapter, we investigate subword units which are better motivated. We also investigate subword-level translation between: (i) related languages that are not very close and (ii) languages do not have a genetic relation, but only a contact relation.

Recently, subword level models have also generated interest in neural machine translation (NMT) systems. The motivation is the need to limit the **vocabulary of neural MT systems**[1] in encoder-decoder architectures (Sutskever et al., 2014). It is in this context that Byte Pair Encoding, a data compression method (Gage, 1994), was adapted to learn subword units for NMT (Sennrich et al., 2016). Other subword units for NMT have also been proposed: character (Chung et al., 2016), Huffman encoding based units (Chitnis and DeNero, 2015), wordpieces (Schuster and Nakajima, 2012; Wu et al., 2016). Our hypothesis is that such subword units learnt from corpora are particularly suited for translation between related languages.

For a more detailed survey on utilizing lexical similarity for MT between related languages, refer to Chapter 2.

3.3 Translation Units for Related Languages

As we have seen, character-level models do not have the representational power to learn mappings between related languages, unless they are very close. Though fixed length character n-gram units increase context size, they are an *ad hoc* choice and increase vocabulary size resulting in data sparsity. Morphemes are variable length units which are linguistically well motivated and have been shown to be useful translation units, but they do not use lexical similarity. Possibly, alternative translation units which can leverage lexical similarity in addition to being variable length and well motivated can help further improve translation quality. We propose the use of two such variable length translation units for SMT between related languages: (a) orthographic syllables, which are linguistically motivated, and (b) byte pair encoded units, which are motivated by the statistical properties of the text. We discuss these translation units in the remainder of the section.

3.3.1 Orthographic syllable

3.3.1.1 From phonemes to syllables

For alphabetic and abugida scripts[2], we can think of characters as the orthographic representation of phonemes. This is a loose correspondence, but useful for the purpose of illustrating the motivation for the translation unit being proposed. Since character units are too short to be useful, we consider orthographic equivalents of longer phonological units. While the phoneme set represents the basic sounds of a language, it is not sufficient to capture the restrictions on the permissible combination of sounds *i.e.*, the phonotactic constraints of the language. Syllables, described by some as phonological *building blocks* of words, are longer phonological units which capture these constraints. Hence, the syllables vocabulary of a language is small and finite. Being

[1] The NMT vocabulary refers to the translation units corresponding to the input and output embeddings.

[2] `https://www.omniglot.com/writing/types.htm`

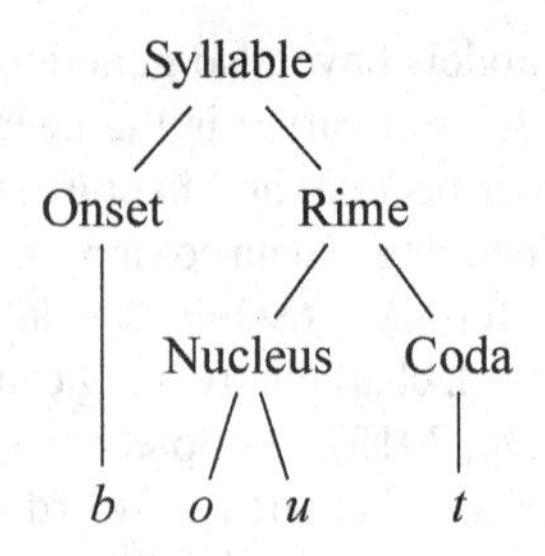

FIGURE 3.1: Structure of a typical syllable illustrated with the English word *bout*.

phonological units, they can also capture lexical similarity across languages. Syllables could act as basic units for translation between related languages.

A syllable is a sequence of speech sounds having one vowel sound, with or without surrounding consonants, forming the whole or a part of a word (Akmajian et al., 2017). The typical syllable consists of two components: an onset followed by a rime. The onset is optional and is a consonant or consonant cluster. The rime can be further divided into the nucleus followed by the coda. The nucleus is mostly mandatory and is a vowel or syllabic consonant. The optional coda is a consonant. Figure 3.1 illustrates the structure of a typical syllable.

Exact syllabification has been widely explored in the literature (Marchand et al., 2009), but it is a non-trivial problem and required phonetic representations. Anyway, the syllable is a phonological unit, and we are interested in its closest orthographic approximation for defining a translation unit. Hence, we explore an approximate syllable unit which we call the *orthographic syllable*. This unit is inspired from the fundamental unit organizational principle of Indic scripts *viz. akshara* (Singh, 2006; Sproat, 2003). Sproat (2003) refers to this unit as an *orthographic syllable*. We generalize the notion of *akshara* beyond Indic scripts. We use Sproat (2003)'s term *orthographic syllable* to refer to this generalized notion also.

Atreya et al. (2016) and Ekbal et al. (2006) have shown that the OS is a useful unit for transliteration involving Indian languages.

3.3.1.2 What is an orthographic syllable?

The *orthographic syllable* is a sequence of one or more consonants followed by a vowel, *i.e* a $C^{+}V$ unit. It can be thought of as an *approximate syllable* with the onset and nucleus, but no coda. This is equivalent to segmenting words at vowel boundaries.

For instance, the English word *spacious* contains the following orthographic syllables (OS) *pre ciou s*. Note that the character sequence *iou* corresponds to a single vowel.

The original notion of *akshara* means a consonant core along with associated vowel diacritics (*maatras*). The consonant core could be single consonant or a consonant cluster (*samkyukta-akshar*). The definition above corresponds to this definition for Indian scripts.

Language	Word	Meaning	OS Segmentation
Marathi	घरासमोर (*gharAsamora*)	in front of the house	घ रा स मो र (*gha rA sa mo ra*)
Hindi	संस्कृति (*sa.mskRRiti*)	culture	सं स्कृ ति (*sa.m skRRi ti*)
Malayalam	സമയം (*samay.m*)	time	സ മ യം (*sa ma y.m*) (*sa ma y.m*)
Tamil	பழம் (*pazham*)	fruit	ப ழ ம் (*pa zha m*)

TABLE 3.1: Examples of orthographic syllables.

For instance, the Hindi word समय (*samaya*) (meaning: *time*) contains the following orthographic syllables: स म य (*sa ma ya*). The corresponding Malayalam loanword സമയം (*samay.m*) contains the following orthographic syllables: സ മ യം (*sa ma y.m*). The two lexically similar words have some common orthographic syllables. Table 3.1 shows a few examples of orthographic syllables for different languages and writing systems.

Since orthographic syllables are identified by vowel boundaries, they can be defined only for writing systems which represent vowels like alphabets or abugidas.

3.3.1.3 Orthographic syllabification

While true syllabification is hard, orthographic syllabification can be easily done with a few rules.

We describe briefly procedures for orthographic syllabification of Indian scripts and non-Indic alphabetic scripts. Orthographic syllabification cannot be done for languages using *logographic* and *abjad* scripts as these scripts do not have vowels.

3.3.1.4 Indic scripts

Indic scripts are *abugida* scripts, consisting of consonant-vowel sequences, with a consonant core (C^+) and a dependent vowel (*matra*). If no vowel follows a consonant, an implicit *schwa* vowel [IPA: ə] is assumed. Suppression of *schwa* is indicated by the *halanta* character following a consonant. This script design makes for a straightforward syllabification process as shown in the following example: लक्ष्मी ($\frac{lakShamI}{CVCCVCV}$) is segmented as ल क्ष मी ($\frac{la}{CV}\ \frac{kSha}{CCV}\ \frac{mI}{CV}$). There are two exceptions to this scheme:

- Indic scripts distinguish between dependent vowels (vowel diacritics) and independent vowels, and the latter will constitute an OS on its own. *e.g.* मुम्बई ($mumbaI$) → मु म्ब ई ($mu\ \ mba\ \ I$). In this case, the last character *I* is an independent vowel.
- The characters *anusvaara* and *chandrabindu* are part of the OS to the left if they represent nasalization of the vowel/consonant or start a new OS if they

Algorithm 1 Orthographic Syllabification of Indic Scripts

```
1: function ORTH-SYLLABIFY(W)
   W: word to be syllabified, SW: syllabified word (spaces added at boundaries)
2:     SW ← ''
3:     for i ← 1, len(W) do
4:         SW ← SW + W[i]
5:         if is_vowel(W[i]) and not is_nasalizer(W[i+1], W[i+2]) then
6:             SW ← SW+' '                                          ▷ Add space
7:         else if is_consonant(W[i]) then
8:             if not ( is_depvowel(W[i+1]) or is_nasalizer(W[i+1], W[i+2])
   ) then
9:                 SW ← SW+' '
10:        else
11:            SW ← SW+' '
12:    return SW

13: function is_nasalizer(c1, c2)
14:     is_nasal ← is_anusvar(c1) or is_chandrabindu(c1)           ▷ Checks for
   certain characters
15:     return is_nasal and not is_plosive(c2)
```

represent a nasal consonant. Their exact role is determined by the character following the *anusvaara*.

Algorithm 1 presents the procedure for orthographic syllabification of Indic scripts.

3.3.1.5 Non-indic alphabetic scripts

We use a simpler method for the alphabetic scripts used in our experiments (Latin and Cyrillic). The OS is identified by a C^+V^+ sequence, *e.g. lakshami→la ksha mi, mumbai→mu mbai.* The OS could contain multiple terminal vowel characters representing long vowels (*oo* in *cool*) or diphthongs (*ai* in *mumbai*). A vowel starting a word is considered to be an OS.

3.3.2 Byte pair encoded unit

3.3.2.1 Beyond syllables

As we have seen, syllables are useful translation units for related languages because they constitute frequent, fundamentals unit for the language. This leads to a small, finite vocabulary which helps reduce data sparsity and utilize lexical similarity. But, we need not restrict ourselves to syllables in order to utilize lexical similarity. There may exist other subword units which capture frequent, subword-level patterns. These frequent subwords need not be linguistic units like syllables, words or morphemes. As phrase-based SMT has shown, it is not necessary that the basic translation units be linguistically valid units to be useful for translation (Koehn et al., 2003). As long as the word *segmentation* captures relevant patterns for which translation mappings can be learnt, it would be useful to consider such translation units.

Moreover, the syllable vocabulary of a language is limited. However, as per Heap's law (Heaps, 1978), the vocabulary of a language tends to increase with an increase in corpus size. This means that subword-level patterns in the data would also increase with corpus size. Syllables will not be sufficient to capture all useful patterns in data. What we need is a procedure to identify subwords patterns that are not necessarily linguistic and that can accommodate an increase in the corpus size.

How do we discover such patterns in text? One solution is to identify the most frequent character subsequences in the text. Byte Pair Encoding (BPE) (Gage, 1994), a text compression algorithm, provides an iterative method to identify frequent subsequences and learn a limited vocabulary. BPE has been used to learn translation units for Neural MT (Sennrich et al., 2016). Another approach proposes to learn the segmentation of the text that maximizes its likelihood (Wu et al., 2016; Schuster and Nakajima, 2012). By learning a compression model for the data, BPE is also implicitly learning to maximize the data likelihood. This interpretation can be understood by taking recourse to the Minimum Description Length (MDL) principle: the best hypothesis (a model and its parameters) for a given set of data is the one that leads to the best compression of the data (Rissanen, 1985). The use of Huffman encoding to discover translation units for NMT (Chitnis and DeNero, 2015) can also be understood in terms of the MDL principle. The vocabulary size is a hyperparameter in these approaches, thus allowing control over the vocabulary size depending on the size, characteristics of the data and the use to which it is put.

These subwords have been proposed for NMT. For NMT, these units enable efficient, high-quality, open vocabulary translation by (i) limiting core vocabulary size, (ii) representing the most frequent words as atomic units and rare words as compositions of the atomic units. These benefits of these subwords are not particular to NMT and apply to SMT between related languages too. Given the lexical similarity between related languages, we would like to *identify a small, core vocabulary of subwords* from which words in the language can be composed.

Given that the above-mentioned subwords are based on the same principle, we study Byte Pair Encoding as a prototypical example of such translation units. As we shall see, BPE units are also easy to learn with a simple algorithm. In the rest of this chapter, we focus on Byte Pair Encoding.

3.3.2.2 Byte pair encoding algorithm

Byte Pair Encoding is a data compression algorithm which was first adapted for Neural Machine Translation by Sennrich et al. (2016) as a way to learn a limited vocabulary for near open vocabulary translation. The essential idea is identification of the most frequent character sequences to add to the initial vocabulary. The input is a monolingual corpus for a language (one side of the parallel training data, in our case). We start with an *initial vocabulary viz.* the characters in the text corpus. The vocabulary is updated using an iterative greedy algorithm. In every iteration, the most frequent bigram (based on current vocabulary) in the corpus is added to the vocabulary (the *merge* operation). Each merge operation adds one new vocabulary item. The corpus is again encoded using the updated vocabulary and this process is repeated for a predetermined number of merge operations. The number of merge operations is the only hyperparameter to the system which needs to be tuned. The final vocabulary size is the sum of the initial vocabulary size (size of the character set) and the number of merge operations.

A new word can be segmented by looking up the learnt vocabulary. For instance, a new word `scion` may be segmented as `sc ion` after looking up the learnt vocabulary, assuming `sc` and `ion` as BPE units learnt during training.

The BPE algorithm does not make any assumptions regarding the nature of the writing system. Hence, BPE segmentation can be performed for text in any writing system.

3.4 Training Subword-level Translation Models

We follow the standard phrase-based SMT training pipeline with some adaptations to account for the constraints imposed by subword-level representation. We describe these adaptations in this section.

3.4.1 Data representation

We segment the data into subwords during pre-processing and indicate word boundaries by a boundary marker (_) as shown in the example below. The boundary marker helps keep track of word boundaries, so the word-level representation can be reconstructed after decoding.

word	`Childhood means simplicity .`
subword	`Chi ldhoo d _ mea ns _ si mpli ci ty _ .`

3.4.2 Language model

We used higher-order language models than typically used with word-level language models. Since the vocabulary size of subword-level models is small, it is viable to train higher-order LMs and estimate n-gram probabilities without being affected by data sparsity (Vilar et al., 2007). Moreover, higher-order LM can help incorporate more context. This could be crucial for subword-level modelling since subword-level translation means lexical mappings use a limited context. Tiedemann (2012a) suggests that highers-order n-grams are crucial to prevent non-word generation for character-level models. We use 10-gram language models for character, OS and BPE-level translation in our experiments.

3.4.3 Tuning

The tuning stage of phrase-based SMT training is concerned with learning the feature weights of the discriminative SMT model. It generally uses an algorithm like Minimum Error Rate Training (MERT) (Och, 2003) or Margin Infused Relaxation Algorithm (MIRA) (Cherry and Foster, 2012) which learns weights that optimize the training objective. Typically, an evaluation metric like word-level BLEU is optimized since it an indicator of the translation quality. Tuning involves a decoder in the loop that is used to generate translation hypothesis for a tuning set. The evaluation metrics for the tuning set are generated, and the feature weights are updated to maximize the score.

For subword-level translation, the decoder generates subword-level output. Since we are interested in optimizing a word-level translation metric like BLEU, we post-process the output of the decoder to generate word-level outputs before score generation and weight updates. Previous work on character-level translation (Nakov and Tiedemann, 2012) has also used the word-level tuning as described here.

3.4.4 Decoding

Since related languages have similar word order, there is hardly any need for reordering of words. Moreover, reordering of words also makes it difficult to ensure that translated subwords of the same source word occur in consecutive positions in the output, complicating the generation of the target language words. In addition, reordering increases the search space exponentially and makes decoding slow. This leads to deterioration in subword-level translation efficiency since the subword representation of sentences is significantly longer than word-level representation. Hence, we choose to disallow reordering during decoding, *i.e.,* perform monotonic decoding.

3.4.5 Desegmentation

It is easy to regenerate words from the segmented output as boundary marker characters indicate word boundaries. Since decoding is monotonic, no reordering of translation units can occur. Hence, we simply concatenate translated subwords between consecutive occurrences of boundary marker characters to reconstruct the translated words.

3.5 Experimental Setup

We trained translation systems over different translation units for comparison: character, morpheme, word, orthographic syllable and BPE unit. In this section, we summarize the languages and writing systems chosen for our experiments and the datasets used. We also describe the experimental configuration of our translation systems and the evaluation methodology.

ben	Bengali	kok	Konkani	pan	Punjabi
bul	Bulgarian	kor	Korean	swe	Swedish
dan	Danish	mac	Macedonian	urd	Urdu
hin	Hindi	mar	Marathi	tam	Tamil
ind	Indonesian	mal	Malayalam	tel	Telugu
jpn	Japanese	may	Malay		

TABLE 3.2: List of languages used in experiments along with ISO 639-3 codes.

Language Family		**Type of Writing System**	
Dravidian	mal,tam,tel	Alphabet	dan[1],swe[1],may[1]
Indo-Aryan	hin,urd,ben		ind[1],buc[2],mac[2]
	kok,mar,pan	Abugida	mal,tam,tel,hin
Slavic	bul,mac		ben,kok,mar,pan
Germanic	dan,swe	Syllabic	kor
Polynesian	may,ind	Logographic	jpn
Altaic	jpn,kor	Abjad	urd

TABLE 3.3: Classification of the languages and writing systems. (i) Indo-Aryan, Slavic and Germanic belong to the larger Indo-European language family. (ii) Alphabetic writing systems used by selected languages: Latin[1] and Cyrillic[2].

3.5.1 Languages and writing systems

Our experiments spanned a diverse set of languages: 16 language pairs, 17 languages and 10 writing systems. Table 3.2 lists the languages and their ISO 639-3 codes. These codes are used in this chapter to refer to these languages. Table 3.3 summarizes the key aspects of the languages involved in the experiments.

The chosen languages span four major language families (six major sub-groups: Indo-Aryan, Slavic and Germanic belong to the larger Indo-European language family). The languages exhibit diversity in word order and morphological complexity. Of course, between related languages, word order and morphological properties are similar. The classification of Japanese and Korean into the Altaic family is debated, but various lexical and grammatical similarities are indisputable, either due to genetic or cognate relationship (Robbeets, 2005; Vovin, 2010). However, the source of lexical similarity is immaterial to the current work. For want of a better classification, we use the name *Altaic* to indicate relatedness between Japanese and Korean.

The chosen language pairs also exhibit varying levels of lexical similarity. Table 3.5 shows an indication of the lexical similarity between them in terms of the Longest Common Subsequence Ratio (LCSR) (Melamed, 1995) (see Section 3.5.4 for information on using LCSR to compute lexical similarity). At one end of the spectrum, Malayalam-India, Urdu-Hindi and Macedonian-Bulgarian are dialects/registers of the same language and exhibit high lexical similarity. At the other end, pairs like Hindi-Malayalam belong to different language families but show many lexical and grammatical similarities due to contact for a long time (Subbārāo, 2012).

The chosen languages cover five types of writing systems. Of these, alphabetic and abugida writing systems represent vowels and logographic writing systems do not have vowels. The use of vowels is optional in abjad writing systems and depends on various factors and conventions. For instance, Urdu word segmentation can be very inconsistent (Durrani and Hussain, 2010) and generally short vowels are not denoted. The Korean *Hangul* writing system is syllabic, so the vowels are implicitly represented in the characters.

3.5.2 Datasets

Table 3.4a shows train, test and tune splits of the parallel corpora used. The Indo-Aryan and Dravidian language parallel corpora are obtained from the multilingual Indian Language Corpora Initiative (ILCI) corpus (Jha, 2012). Parallel corpora for other pairs were obtained from the *OpenSubtitles2016* section of the OPUS corpus collection (Tiedemann, 2012b). Language models for word-level systems were trained on the target side of training corpora plus additional monolingual corpora from various sources (see Table 3.4b for details). We used just the target language side of the parallel corpora for character, morpheme, OS and BPE-level LMs.

3.5.3 System details

3.5.3.1 Phrase-based SMT

We trained phrase-based SMT systems using the *Moses* system (Koehn et al., 2007), with the *grow-diag-final-and* heuristic for symmetrization of word alignments, and Batch MIRA (Cherry and Foster, 2012) for tuning. Subword-level representation increases the sentence length, considerably impacting decoding and tuning speed. Hence, we speed up decoding by using cube pruning with a smaller stack size (pop-limit = 1000). In the next chapter, we present a study on the effect of decoder settings and show that this setting has minimal impact on translation quality while reducing decoding and tuning time. We trained 5-gram LMs with Kneser-Ney smoothing for word and morpheme-level models and 10-gram LMs for the character, OS and BPE-level models using SRILM (Stolcke et al., 2002).

3.5.3.2 Morpheme segmentation

We used unsupervised morphological-segmenters for generating morpheme representations for Indian languages since morphological analyzers are not available for many languages. The morphological segmenters were trained using *Morfessor 2.0*[3] (Smit et al., 2014). The morphanalyzers are learnt from monolingual corpora using a probabilistic generative model which uses maximum-a-posteriori estimation with sparse priors inspired by the Minimum Description Length (MDL) principle (Virpioja et al., 2013). Even though these morphanalyzers do not explicitly model stem, prefix, suffix or provide any grammatical properties, the segmentation generated by

[3] http://www.cis.hut.fi/projects/morpho/morfessor2.shtml

Language Pair	Train	Tune	Test
ben-hin, pan-hin, kok-mar, mal-tam, tel-mal, hin-mal, mal-hin	44,777	1000	2000
urd-hin, ben-urd urd-mal, mal-urd	38,162	843	1707
bul-mac	150k	1000	2000
dan-swe	150k	1000	2000
may-ind	137k	1000	2000
kor-jpn,jpn-kor	69,809	1000	2000

(a) Parallel Corpora

Language	Size
hin (Bojar et al., 2014)	10M
ben (Goldhahn et al., 2012)	400K
ind (Tiedemann, 2012b)	640K
mac (Tiedemann, 2012b)	680K
mal (Goldhahn et al., 2012)	200K
mar (news websites)	1.8M
swe (Tiedemann, 2012b)	2.4M
tel (Goldhahn et al., 2012)	600K
tam (Ramasamy et al., 2012)	1M
urd (Jawaid et al., 2014)	5M

(b) Monolingual corpora

TABLE 3.4: Training corpus statistics (in number of sentences).

this method is sufficient for preprocessing parallel corpora. We used only the word types without considering their frequencies for training since this training configuration has been shown to perform better when no annotated data is available for tuning (Virpioja et al., 2013).

We used *Juman*[4] and *Mecab*[5] for tokenization and morphanalysis of Japanese and Korean, respectively.

3.5.3.3 OS and BPE segmentation

We implemented the orthographic syllabification rules discussed in the previous section. For training BPE models, we used the *subword-nmt*[6] library.

[4]http://nlp.ist.i.kyoto-u.ac.jp/EN/index.php?JUMAN

[5]https://bitbucket.org/eunjeon/mecab-ko

[6]https://github.com/rsennrich/subword-nmt

3.5.4 Evaluation

3.5.4.1 BLEU

The primary evaluation metric is word-level BLEU (Papineni et al., 2002). BLEU relies on an exact match of n-grams between the reference and hypothesis sentences. This is problematic for morphologically rich languages (highly inflected or compounding languages) since exact matches may be hard to come by. It is also problematic for evaluating subword-level translation models, since even a difference of a single character between the reference and hypothesis words will result in a missed match.

3.5.4.2 LeBLEU

Hence, we also used a secondary evaluation metric, Levenshtein Edit BLEU (LeBLEU) (Virpioja and Grönroos, 2015), which is more suitable for morphologically rich languages. LeBLEU is a variant of BLEU that does soft-matching of words based on edit distance and has been shown to be better for morphologically rich languages.

3.5.4.3 LCSR

We use the Longest Common Subsequence Ratio (LCSR) (Melamed, 1995) as a diagnostic metric for different purposes. Given two strings x and y, the LCSR between the two strings can be computed as:

$$\text{LCSR}(x, y) = \frac{\text{LCS}(x, y)}{\max(len(x), len(y))}, \tag{3.1}$$

where, LCS is the longest common subsequence between x and y. We consider the character as the basic unit for computation of LCSR since we are interested in subword-level correspondences for related languages. LCSR is a versatile metric and can be used in different ways:

- The LCSR between two words is a measure of the lexical similarity between the two words.
- The scope of LCSR computation can be extended to an entire sentence. In that case, character-level LCSR between two sentences in a language is a measure of lexical similarity between the two sentences. Sentence LCSR between the output of an MT system and the reference translation can be considered a proxy measure for sentence-level translation quality. It is well correlated with BLEU as well as LeBLEU.
- The LCSR between a parallel sentence pair in two languages quantifies the cross-lingual lexical similarity between the two sentences. In order to compute LCSR across two languages, the languages must use the same or compatible

writing systems. By 'compatible writing systems', we mean that a one-one correspondence can be established between most characters in the writing systems. *e.g.*, (1) Latin and Cyrillic writing systems, (2) Indic writing systems derived from the Brahmi script.

- Given a representative parallel corpus ($\mathcal{D}$) in languages L_1 and L_2, we can estimate the lexical similarity between the two languages by the mean LCSR over the sentence pairs in the parallel corpus.

$$sim(L_1, L_2) = \frac{1}{|\mathcal{D}|} \sum_{(x,y)\in\mathcal{D}} \text{LCSR}(x, y) \tag{3.2}$$

3.5.4.4 Bootstrap resampling

We used bootstrap resampling to establish if the differences in BLEU scores in our experiments are statistically significant (Koehn, 2004).

3.6 Results and Discussion

In this section, we report and discuss the results of experiments. These experiments investigate the following questions:

- Are BPE and OS units better than other units for translation between related languages?
- How do BPE and OS-level translation compare with translation systems which transliterate untranslated words to improve word and morpheme-level baseline systems?
- The length of sentences (in terms of translation units) depends on the choice of the translation unit. Does the sentence length have an impact on translation quality?

3.6.1 Comparison between different translation units

Table 3.5 shows BLEU scores for all the language pairs and translation units we experimented with, along with lexical similarity between the language pairs (in terms of LCSR) .

A few notes about the experimental configuration for results reported in the table:

- *Number of BPE merge operations*: To learn the BPE vocabulary, we have to apriori specify the number of BPE merge operations, which is a hyperparameter to the BPE algorithm. We choose the number of BPE merge operations to

Src-Tgt	LCSR	Char	Word	Morph	OS	BPE
ben-hin	52.30	27.95	32.47	32.17	**33.54**	33.22
pan-hin	67.99	71.26	70.07	71.29	**72.41**	72.22
kok-mar	54.51	19.83	21.30	22.81	23.43	**23.63**
mal-tam	39.04	4.50	6.38	7.61	7.84	**8.67**†
tel-mal	39.18	6.00	6.78	7.86	8.50	**8.79**
hin-mal	33.24	6.28	8.55	9.23	10.46	**10.73**
mal-hin	33.24	12.33	15.18	17.08	18.44	**20.54**
urd-hin	-	52.57	55.12	52.87	NA	**55.55**
ben-urd	-	18.16	27.06	27.31	NA	**28.06**
urd-mal	-	3.13	6.49	7.05	NA	**8.44**
mal-urd	-	8.90	13.22	15.30	NA	**18.48**
bul-mac	62.85	20.61	21.20	-	**21.95**	21.73
dan-swe	63.39	35.36	35.13	-	35.46	**35.77**
may-ind	73.54	60.50	**61.33**	-	60.79†	59.54
kor-jpn	-	8.51	9.90	-	NA	**10.23**
jpn-kor	-	8.17	8.44	-	NA	**9.02**

TABLE 3.5: Translation quality for various translation units (BLEU). The values marked in **bold** indicate best score for a language pair. † indicates that difference in BLEU scores between BPE and OS are statistically significant ($p < 0.05$).

ensure that the vocabulary size of the resultant BPE-level training set is equivalent to the vocabulary size of the OS-level training set. This choice allows a fair comparison between OS and BPE-level models. Of course, the BPE vocabulary size may have an impact on translation quality; hence, we study the effect of the number of BPE merge operations in the next chapter.

Since orthographic syllabification cannot be performed for Urdu, Korean and Japanese, we choose the number of BPE merge operations as follows:

- *Urdu*: The number of BPE merge operations was selected based on the Hindi OS vocabulary since Hindi and Urdu are registers of the same language.
- *Korean and Japanese*: We choose 3000 BPE merge operations based on searching for the number of BPE operations that gave the best translation quality on a separate validation set.

- *Morphological Analysis for Japanese and Korean*: Since Japanese and Korean texts are not segmented, we used off-the-shelf tools for tokenization, which implicitly also perform morphological segmentation. Hence, we do not distinguish between word and morpheme-level for these languages. The results are reported under the word-level column.
- We focus on Indic languages for qualitative analysis, since the authors are familiar with them and they show diversity in morphological richness as well

as language relatedness. Most of these languages are genuinely low-resource languages and Indic languages are a motivating factor for this monograph.

Our major observations for each unit of translation are described in the following sections.

3.6.1.1 Word and morpheme

We see that morpheme-level translation outperforms word-level translation for most language pairs. The differences in BLEU scores are minor for translation involving Indo-Aryan languages. However, we see major improvements for translation involving Dravidian languages (about 13% improvement in BLEU score). The improvements are particularly impressive when both source and target languages are Dravidian languages (about 18% improvement in BLEU score). Dravidian languages are highly agglutinative, while Indo-Aryan languages have a relatively less synthetic morphology. Therefore, morphological segmentation benefits translation involving languages with a synthetic morphology.

Segmenting a word into its constituent morphemes reduces the number of out-of-vocabulary (OOV) words and mitigates data sparsity to a certain extent. Between related languages, the system can learn mappings between isomorphic forms of morphemes across languages. For instance, it can learn to mappings of suffixes/post-positions as illustrated below for Hindi-Marathi:

Hindi	Marathi	Meaning
बिना (*binA*)	शिवाय (*shivAya*)	without
बारे में (*bAre me.n*)	बाबतीत (*bAbatIta*)	about
जैसा (*jaisA*)	सारखा (*sArakhA*)	like (preposition)

However, morpheme-level translation does not leverage lexical similarity, nor does word-level translation.

3.6.1.2 Character

The simplest translation unit that can utilize lexical similarity is the character. However, the character-level model is competitive with the word-level model only for *very close* language pairs with high lexical similarity like Punjabi-Hindi, Urdu-Hindi, Konkani-Marathi, Bulgarian-Macedonian, Danish-Swedish, Malay-Indonesian. The BLEU scores of the character-level models are within 5% of the word-level models' scores for these language pairs. This is consistent with observations made in the literature (Nakov and Tiedemann, 2012), where character-level models have been used for translation between *very close* languages. Even in this case, the character-level models did not consistently outperform word-level models.

Moreover, character-level models are far inferior to word-level models for language pairs that have a relatively lower lexical similarity (average BLEU scores are 25% lower). This shows that character-level models are not effective for translation between related languages with moderate lexical similarity. We need better translation units that harness the lexical similarity between such language pairs.

As pointed out earlier, the small context of unigram character-level models is a major limitation. Hence, we explored trigram character-level translation models, but the results were worse than unigram character-level models. The trigram models increase the context size for translation, but the vocabulary size also increases significantly. Data sparsity again becomes a concern and hence trigram-level models provide no improvement in translation. It is clear that *ad hoc* increase in context size is not beneficial for translation. What we need is a principled way of identifying relevant translation units.

3.6.1.3 Orthographic syllable

The orthographic syllable-level system is clearly better than the character, morpheme and word-level translations.

It significantly outperforms the character-level model (by 38.9% for Indic language pairs using *abugida* scripts). Clearly, OS-level translation makes better use of lexical similarity compared to character-level translation. The longer, variable length units are a better representation for translation between related languages.

OS-level translation outperforms the word- and morpheme-level models by 15.5% and 5.9%, respectively, for Indic language pairs using *abugida* scripts. The OS-level models are more beneficial for the more synthetic Dravidian languages. They show an improvement of 23% and 8% over word- and morpheme-level models, respectively, for language pairs involving Dravidian languages.

OS-level translation also outperforms word and morpheme-level models for translation between language pairs belonging to different language families, but having a long contact relationship *viz.* Malayalam-Hindi and Hindi-Malayalam. Thus, OS-level translation can make use of lexical similarity arising as a result of contact relation among Indo-Aryan and Dravidian languages also.

Few orthographic syllables could not be translated (OOV orthographic syllables). We handled them by simple source to target script mapping of characters in the untranslated syllables. This barely affected the translation quality (0.02% increase in BLEU score).

3.6.1.4 Byte pair encoded unit

BPE units are clearly better than the traditional word and morpheme representations. We observe an average BLEU score improvement of 15% and 11% over word-level and morpheme-level translation, respectively. The only exception is Malay-Indonesian, which are registers of the same language.

BPE units also show modest improvement over orthographic syllables for most language pairs (average improvement of 2.6% and maximum improvement of up to 11%). The improvements are not statistically significant for most language pairs. The only exceptions are Bengali-Hindi, Punjabi-Hindi and Malay-Indonesian – all these languages pairs have relatively less morphological affixing (Bengali-Hindi, Punjabi-Hindi) or are registers of the same language (Malay-Indonesian). For Bengali-Hindi and Punjabi-Hindi, the BPE-level BLEU scores are quite close to OS-level BLEU scores.

It is worth mentioning that BPE-level translation provides a substantial improvement over OS-level translation when a morphologically rich language is involved. In translations involving Dravidian languages, we observe 6.25% average increase in the BLEU score.

Like OS-level translation, BPE-level translation also outperforms other units for translation between language families belonging to different language pairs, but having a long contact relationship *viz.* Malayalam-Hindi and Hindi-Malayalam.

3.6.1.5 Applicability to different writing systems

The utility of orthographic syllables as translation units is limited to languages that use writing systems which represent vowels. Alphabetic (Roman, Cyrillic) and abugida (Indic scripts) writing systems fall into this category. On the other hand, logographic writing systems (Japanese Kanji, Chinese) and abjad writing systems (Arabic, Hebrew, Syriac) do not represent vowels. To be more precise, abjad writing systems may represent some/all vowels depending on the language, pragmatics and conventions. Syllabic writing systems like Korean Hangul do not explicitly represent vowels since the basic unit (the syllable) implicitly represents the vowels.

The major advantage of Byte Pair Encoding is its **writing system independence** and our results show that BPE encoded units are useful for translation involving abjad (Urdu uses an extended Arabic writing system), logographic (Japanese Kanji) and syllabic (Korean Hangul) writing systems as well. For language pairs involving Urdu, there is an 18% average improvement over word-level and 12% average improvement over morpheme-level translation quality. For Japanese-Korean language pairs, an average improvement of 6% in BLEU scores over a word-level baseline is observed.

3.6.2 Evaluation using additional translation quality metrics

We have seen that subword-level translation models improve translation quality as measured by the BLEU metric. However, BLEU has a few limitations and does not shed any light on the nature of the improvements. Hence, we evaluate the translation output using additional metrics that provide insights regarding improvement in translation quality.

3.6.2.1 LeBLEU

BLEU has many limitations when used for evaluation of translation involving morphologically rich languages. Since our experiments involve morphologically rich languages, we also evaluate our systems using the LeBLEU metric (Virpioja and Grönroos, 2015), a metric suited for morphologically rich languages. Table 3.6 shows the LeBLEU scores. We see that the LeBLEU scores show the same trends as BLEU scores *i.e.,* BPE and OS units outperform other translation units. The only exception is the case of very close languages like Danish-Swedish and Malay-Indonesian,

Src-Tgt	LCSR	Char	Word	Morph	OS	BPE
ben-hin	52.30	0.672	0.682	0.708	0.715	**0.716**
pan-hin	67.99	0.905	0.871	0.899	0.906	**0.907**
kok-mar	54.51	0.632	0.636	0.659	**0.671**	0.665
mal-tam	39.04	0.311	0.314	0.409	0.447	**0.465**
tel-mal	39.18	0.346	0.314	0.383	0.439	**0.443**
hin-mal	33.24	0.324	0.393	0.436	**0.477**	0.468
mal-hin	33.24	0.444	0.460	0.528	0.551	**0.565**
urd-hin	-	0.804	0.795	0.792	NA	**0.823**
ben-urd	-	0.607	0.660	0.671	NA	**0.692**
urd-mal	-	0.247	0.350	0.379	NA	**0.416**
mal-urd	-	0.444	0.454	0.522	NA	**0.568**
bul-mac	62.85	0.603	0.606	-	**0.613**	0.599
dan-swe	63.39	0.692	**0.694**	-	0.682	0.682
may-ind	73.54	0.827	**0.832**	-	0.828	0.825
kor-jpn	-	0.396	0.372	-	NA	**0.408**
jpn-kor	-	0.372	0.350	-	NA	**0.374**

TABLE 3.6: Translation quality for various translation units (LeBLEU).

where the word-level translation seems to be performing well as per LeBLEU metrics. Thus, the effectiveness of subword-level translation is validated by a translation metric more appropriate to subword-level translation too. In the remainder of this chapter, we report just BLEU scores for the sake of brevity.

3.6.2.2 Recall and precision

Table 3.7 shows the unigram recall and precision of the translation output with respect to the reference translation[7]. We observe a significant improvement in recall for OS and BPE-level translation over word-level translation, suggesting that the system is able to correctly translate out-of-vocabulary words. A similar improvement is observed for precision too. We do not observe any major gains in recall over morpheme-level translation, but observe modest improvement in precision.

3.6.2.3 Effect of sentence length on translation quality

Subword-level models increase the length of the sentence to be translated by the decoder (in terms of the number of subword units). Increased length results in a large search-space for the decoder and potentially more search errors. Does the increase in length adversely affect translation quality? To study the impact of sentence length on translation quality, we study the Pearson's correlation coefficient between (1) source

[7]We created a custom version of the METEOR tool (Banerjee and Lavie, 2005) for Indian languages, called *METEOR-Indic*, to compute recall and precision: `https://github.com/anoopkunchukuttan/meteor_indic`

Src-Tgt	Char	Word	Morph	OS	BPE
ben-hin	0.633	0.668	0.678	**0.680**	**0.680**
pan-hin	0.884	0.871	0.881	**0.887**	**0.887**
kok-mar	0.551	0.567	0.589	0.589	**0.591**
mal-tal	0.242	0.299	0.336	0.321	**0.347**
tel-mal	0.297	0.327	0.354	0.354	**0.361**
hin-mal	0.288	0.373	0.377	0.377	**0.388**
mal-hin	0.418	0.450	0.505	0.514	**0.542**

(a) Recall

Src-Tgt	Char	Word	Morph	OS	BPE
ben-hin	0.652	0.673	0.676	0.686	**0.689**
pan-hin	0.887	0.881	0.888	0.895	**0.894**
kok-mar	0.570	0.576	0.587	0.599	**0.600**
mal-tal	0.300	0.324	0.340	0.347	**0.362**
tel-mal	0.295	0.312	0.340	0.348	**0.353**
hin-mal	0.288	0.340	0.360	0.366	**0.380**
mal-hin	0.481	0.503	0.516	0.542	**0.565**

(b) Precision

TABLE 3.7: Recall and precision computed using *METEOR-Indic*.

sentence length (in terms of the number of words) and (2) sentence-level translation quality. We use LCSR at the character-level for computing the translation quality. Figure 3.2 depicts this correlation for different language pairs and translation units. We see that there is no correlation between sentence length and translation quality. However, the increase in sentence length substantially increases the decoding time. In the next chapter, we look at solutions to speed up decoding.

3.6.3 Comparison with transliteration post-processing

We also compare OS and BPE-level models with another approach which utilizes lexical similarity *viz. transliteration post-processing*.

Words that could not be translated by the word-level system are transliterated during post-processing. For transliteration, we use a statistical machine transliteration model based on phrase-based SMT (Rama and Gali, 2009). The transliteration model was trained on parallel transliteration corpora mined from the parallel translation corpus we used for training the translation models. The transliteration corpus was mined using an EM-based unsupervised transliteration mining algorithm (Sajjad et al., 2012). Since this method relies on lexical similarity and uses the parallel translation corpus for mining transliterations, the mined corpus contains named entities, cognates, loanwords and lateral borrowings. Therefore, the mined transliteration corpus is a good representative of the lexical changes between the related source and target languages.

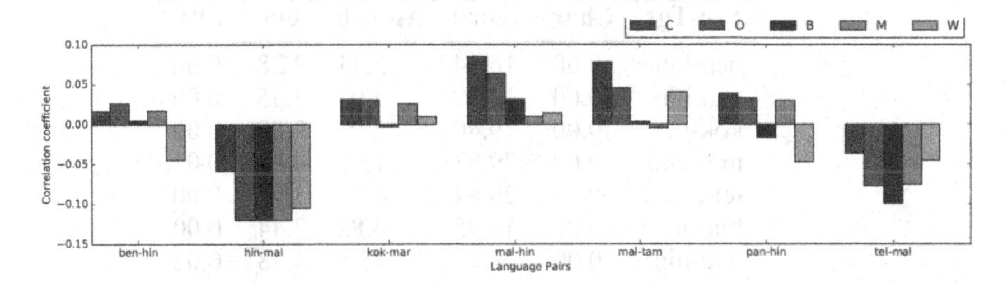

FIGURE 3.2: Pearson's correlation coefficient between source sentence length and translation quality. (C: character, M: morpheme, W: word, O: orthographic syllable, B: BPE).

Src-Tgt	Word$_X$	Morph$_X$	OS	BPE
ben-hin	32.79	32.32	**33.54**	33.22
pan-hin	71.71	71.42	**72.41**	72.22
kok-mar	21.9	22.82	23.43	**23.63**
mal-tam	7.01	7.65	7.84	**8.67**
tel-mal	6.94	7.89	8.50	**8.79**
hin-mal	8.77	9.26	10.46	**10.73**
mal-hin	16.26	17.3	18.44	**20.54**

TABLE 3.8: Comparison of transliteration post-processing with subword-level translation (BLEU).

Hence, word-level translation followed by transliteration post-processing forms a stronger baseline compared to the vanilla word-level translation system (we refer to this system as *Word*$_X$). We also compare with transliteration post-processing of morpheme-level translation (we refer to this system as *Morph*$_X$).

Table 3.8 shows a comparison of the transliteration post-processing approach with OS and BPE-level models described earlier. The OS-level models show an improvement of 10% and 5% in BLEU score over the *Word*$_X$ and *Morph*$_X$ systems, respectively. The BPE-level models show an improvement of 15% and 9% in BLEU score over the *Word*$_X$ and *Morph*$_X$ systems, respectively. Thus, OS and BPE-level models significantly outperform these stronger transliteration post-processing baselines too.

3.7 Why Are Subword Units Better Than Other Translation Units?

In this section, we discuss the reasons for the improved performance of subword units over other translation units:

Src-Tgt	Char	Word	Morph	OS	BPE
ben-hin	0.00	16.14	3.31	2.28	0.00
pan-hin	0.00	14.09	4.03	3.35	0.00
kok-mar	0.00	19.40	2.92	2.89	0.00
mal-tam	0.00	29.83	13.5	4.38	0.02
tel-mal	0.00	26.84	6.32	3.97	0.00
hin-mal	0.00	11.95	4.82	2.44	0.00
mal-hin	0.00	29.83	13.5	4.38	0.02

TABLE 3.9: Source OOV rate (for translation unit types).

3.7.1 Reduction in data sparsity

One of the major challenges of working with limited corpora is data sparsity. At the word level, the training corpus captures only a fraction of the language vocabulary. Hence, the translation system encounters a significant number of OOV words at decoding time. We can quantify the scale of the OOV problem by the source OOV type rate *i.e.,* the fraction of the word-types in the source language test set not seen in the corresponding training set as the ratio of the number of word-types in the training set.

$$\text{OOV}_{\text{rate}} = \frac{|\mathcal{V}_{\text{test}} - \mathcal{V}_{\text{train}}|}{|\mathcal{V}_{\text{test}}|} \tag{3.3}$$

where, $\mathcal{V}_{\text{train}}$ and $\mathcal{V}_{\text{test}}$ are the sets of word-types in the training and test set respectively.

We can define source OOV rates for other translation units also similarly. Table 3.9 shows the source OOV rates for different translation units and language pairs. We can see that the OOV rate is very high for word-level representation, especially when the source language is a Dravidian language. The OOV rate reduces progressively for morpheme, OS and BPE-level representations. In fact, the OOV rates are low for OS and negligible for BPE and character-level representations.

To study the impact of OOV rate on translation quality, we study the Pearson's correlation coefficient between: (1) ratio of OOV words to total words in the source sentence at sentence level and (2) sentence-level translation quality. We use LCSR at the character-level for computing the translation quality. Figure 3.3 depicts this correlation for different language pairs and translation units. We see that word-level models show moderate negative correlation between source OOV rate and translation quality. For morpheme-level models, we see a moderate negative correlation only when the source language is morphologically rich. On the other hand, there is no correlation for BPE, OS and character-level models since these models have very few OOVs. Hence, subword-level representation reduces the adverse effects of OOVs on translation quality.

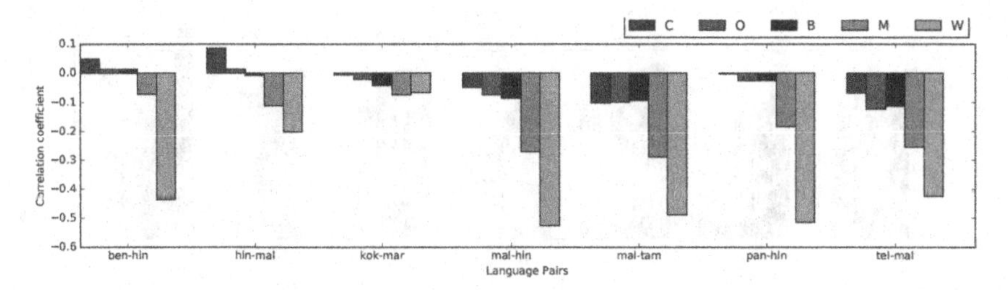

FIGURE 3.3: Pearson's correlation coefficient between source OOV rate (for word-type) and translation quality.

3.7.2 Judicious use of lexical similarity

But character-level models also have a limited vocabulary, yet they do not improve translation quality except for very close languages. Character-level models learn character mappings effectively, which is sufficient for translating related languages which are very close to each other (translation is akin to transliteration in these cases). But this is insufficient for translating related languages that are more divergent since translating morphological affixes, lexically dissimilar words, some lexically similar words *etc.* requires a larger context. Character-level models do not have the representational power to model these translation phenomena. It seems that translation quality is highly correlated to lexical similarity between the source and target sentences. On the other hand, words and morphemes do not depend on lexical similarity. Where do OS and BPE-levels models stand with respect to use of lexical similarity?

To understand this, we study the Pearson's correlation coefficient between (1) lexical similarity of the reference sentence pair (in the test set) and (2) sentence-level translation quality. We use LCSR at the character-level for computing the source/target lexical similarity and the translation quality. Figure 3.4a depicts this correlation for different language pairs and translation units. As hypothesized, character-level models show the highest correlation between translation quality and lexical similarity. We see that character-level models work best when the source and target sentences have high lexical similarity but perform badly otherwise. Not surprisingly, word-level models show the least correlation, followed by morpheme-level models since both of them do not utilize lexical similarity.

If we look at OS and BPE-level models, they exhibit lesser correlation than character-level models and are able to make **generalizations beyond simple character-level mappings**. They sit somewhere between character-level and word/morpheme-level systems in the correlation spectrum. Thus, these subword models maintain a judicious balance between learning lexical similarity based mappings and higher-level morpheme and word-level mappings.

As we have seen, character-level models are not powerful enough to exploit lexical similarities. However, the OS and BPE models are better at utilizing lexical similarity than character-level models. Their translation outputs are more lexically similar to the source sentences than the character-level models. This can be seen from a study

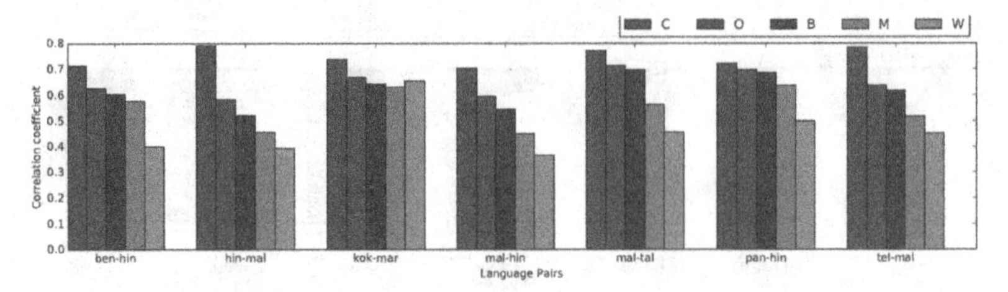

(a) Correlation between lexical similarity and translation quality

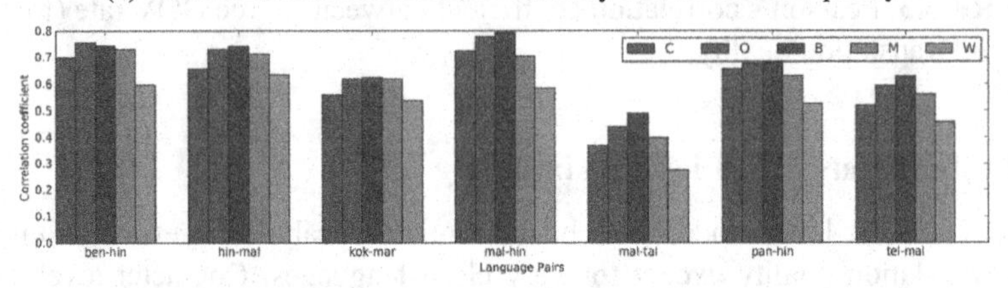

(b) Correlation between lexical similarity of (i) source-reference target and (ii) source-system output

FIGURE 3.4: Correlation studies involving lexical similarity (uses Pearson's coefficient).

	hin	**mar**	**mal**
OS	tI, stha	mA, nA	kka, nI
Suffix	ke, me.m	ChyA, madhIla	unnu, .e∼Nkill.m
Word	paryaTaka, athavA	prAchIna, aneka	bhakShaN.m, yAtra

TABLE 3.10: Examples of BPE units for Indian languages.

of the correlation between the following lexical similarities of: (1) source and reference target sentences, (2) source and output target sentences. Figure 3.4b depicts this correlation for different language pairs and translation units. OS and BPE units show the highest correlation between the lexical similarities, indicating they are able to utilize lexical similarity better than other translation units.

3.7.3 Ability to learn diverse lexical mappings

Lexical similarity cannot be modelled by an *ad hoc* choice of subword translation units. We have seen that character-level models do not have the representational power to model translation between related languages. Character n-gram models run into data sparsity problems. Hence, a principled approach, that can model lexical similarity as well as semantic similarity while limiting the vocabulary size, is necessary for selecting the subword vocabulary.

As opposed to arbitrary n-gram units, the BPE and OS units are well motivated. Orthographic syllables represent approximate syllables. The syllable inventory of a

Source	Target	Meaning
വാതിൽ (*vAtil*) (*mal*)	दर्वाजा (*darvAaA*) (*hin*)	door
മുമ്പ് (*mump*) (*mal*)	पहले (*pahale*) (*hin*)	ago

TABLE 3.11: Examples of mappings of non-cognates learnt by BPE and OS-level translation for different language pairs.

language is finite and small. Through the process of identifying variable length frequent subsequences, we observe that BPE units also **represent higher-level semantic units like frequent morphemes, suffixes and entire words**. Table 3.10 shows a few examples for some Indian languages. So, BPE-level models can learn semantically similar translation mappings in addition to lexically similar mappings. In this way, BPE units enable translation models to **balance utilization of lexical similarity and semantic similarity**.

Both OS and BPE-level models are able to correctly translate words whose translations are not lexically similar to their source language synonyms. Table 3.11 illustrates a few such examples. This is an important property since: (i) function words and suffixes tend to be less similar lexically across languages and (ii) there is a fair share of non-lexically similar words between two related languages also. The alignment template driven phrase extraction approach used by PBSMT probably allows learning of such non-cognate mappings also.

BPE units also have an **additional degree of freedom** (choice of vocabulary size), which allows tuning for best translation quality.

3.8 Summary and Future Work

In this chapter, we explore methods to improve subword-level translation in order to effectively utilize lexical similarity between related languages. To this end, we propose the use of two subword units for translation between related languages: orthographic syllables and byte pair encoded units. Unlike character n-grams, which are ad hoc translation units employed to utilize lexical similarity, our proposed units are well-motivated. While orthographic syllables are linguistically motivated pseudo-syllables, BPE identifies the most frequent character sequences as basic units based on statistical properties of the text. The following are the major findings:

- We show that both these translation units significantly outperform previously proposed translation units *viz.* characters, character n-grams, morphemes and words. They also outperform a strong baseline which transliterates untranslated words in the output of word- and morpheme-level translation models. Our results are supported by extensive experimentation spanning multiple language families and writing systems.

- We show that the proposed subword translation units can leverage lexical similarity for languages that share genetic and/or contact relationship. Previous work using character-level translation had shown improvement only for languages sharing a genetic relationship. Moreover, character-level models work only in restricted scenarios where the languages involved are *very close* (mutually intelligible languages, dialects, registers, *etc.*). OS and BPE units perform well under far more relaxed conditions too.

- While orthographic syllables can be used only for languages whose writing systems use vowel representations, BPE units are writing system independent and perform well for such languages too. BPE also enables discovery of translation mappings at various levels simultaneously (syllables, suffixes, morphemes, words, *etc.*). For both units, the translation models are able to learn not just cognate mappings, but non-cognates as well. The translation quality of OS and BPE-level models are comparable. However, BPE-level models have an additional degree of freedom - the ability to tune the vocabulary size. We will discuss this notion in the next chapter.

- The improvement in translation quality due to OS and BPE units can be attributed to the following reasons: (i) reduction in data sparsity, (ii) judicious use of lexical similarity and (iii) ability to learn diverse lexical mappings. We provide quantitative and qualitative observations in support of this analysis.

Given the improved translation quality due to these subword-level models, it would be natural to apply subword-level models for translation-related tasks involving related languages *viz.* pivot-based MT, domain adaptation (Tiedemann, 2012a) and translation between a *lingua franca* and related languages (Wang et al., 2012). We explore cross-domain translation in the next chapter and pivot-based MT in Chapter 5.

In the next chapter, we continue our investigations with subword-level translation models with a view to further improving translation quality. We are particularly interested in investigating how subword-level MT performs in resource-scarce scenarios. We also investigate different design choices in building subword-level translation systems.

Chapter 4

Improving Subword-level Translation Quality

In the previous chapter, we proposed various subword units that can improve translation between related languages by utilizing lexical similarity. We empirically showed that subword units are indeed beneficial and analyzed the reasons for the improved performance due to subword units. This chapter continues our investigation of subword units for translation between related languages. In this chapter, we focus on improving the performance of subword-level translation. Broadly, the investigation is along the following lines:

- As a major motivation for subword translation is improving translation quality under low-resource scenarios, we study how translation quality is affected by the availability of resources, *viz.* parallel corpora, for different translation units. (Section 4.1)
- We discussed the overall algorithm for training subword-level models in the previous chapter. However, a number of design choices and hyperparameter settings are key to getting the best performance from subword-level translation. We present an empirical study of these design choices and hyperparameter settings in this chapter. (Section 4.2)
- While translation quality is the most important metric for system's performance, the system should be able to translate at a reasonably good speed for it to be practically useful. Subword-level representation increases the sentence length (in terms of translation units), which could affect decoding speed. Hence, we study factors affecting decoding speed and suggest optimizations to achieve acceptable decoding speed. (Section 4.3)

4.1 Effect of Resource Availability

In this section, we present a study of the effect of parallel corpora availability for different translation units. We study two variable factors related to parallel corpus availability: (i) the size of the parallel corpus and (ii) change of translation domain.

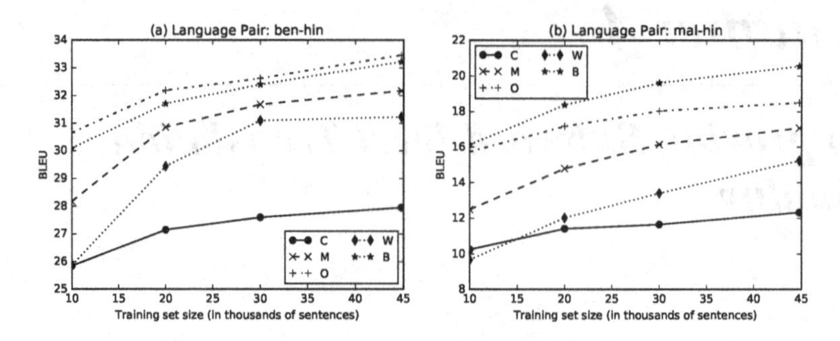

FIGURE 4.1: Effect of training data size on translation quality for different translation units (C: character, M: morpheme, W: word, O: Orthographic syllable, B: BPE).

4.1.1 Effect of parallel corpus size

As we have seen earlier, subword units mitigate the problem of data sparsity, thus improving translation quality. We would, therefore, expect that subword-level translation models can perform well with smaller parallel corpora. On the other hand, word-level models have traditionally benefitted from large corpora on account of lexical acquisition. Given the limited vocabulary of subword-level models, it is pertinent to ask if subword-level models also benefit from large parallel corpora.

To investigate these questions, we trained translation models with different translation units for different training sizes. We experimented with the following training set sizes: 10K, 20K, 30K, 44,777 sentences. We used the same experimental settings as described in Chapter 3. Figure 4.1 shows the learning curves for translation between Bengali-Hindi and Malayalam-Hindi.

The following are the major observations:

- Character-level models do not show any significant improvement in translation quality with increasing parallel corpus size. More parallel corpora would most likely not bring any major gains in translation quality. This further underscores the observations in the previous chapter and shows that character-level models do not have the representational power for translation between related languages.

- Word-level models clearly benefit from an increased training corpus. They perform very poorly on very small corpora, but make rapid gains in translation quality as corpus size increases.

- Morpheme-level models trained on small training corpora are better than the corresponding word and character-level models. In fact, morpheme-level models trained on about 10K sentence pairs are competitive with character-level models trained more than 40K sentences. Translation quality increases steadily with increase in corpus size.

- Both OS and BPE-level models perform reasonably well on very small training corpora. The translation quality on 10K sentence pairs approaches the

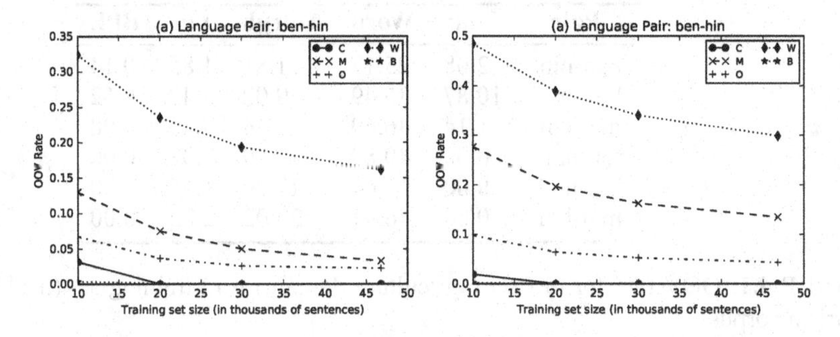

FIGURE 4.2: Effect of training data size on OOV rate for different translation units.

translation quality of the word-level system on the entire training set. With about 20K sentence pairs, the translation quality approaches the accuracy of the morpheme-level system on the entire corpus. Note that Malayalam and Hindi belong to different language families. Still, these models are able to learn reasonably good translation models from small parallel corpora by leveraging lexical similarity.

- OS and BPE-level models consistently outperform other translation units for all training corpora sizes. BPE-level models perform better than OS-level models.
- Except for character, models trained on other units do not seem to have saturated after training on the entire corpus. Further increase in training corpora should bring more improvement for each of these models. It remains to be investigated which models saturate faster with an increase in training corpora. The OOV rate seems to be an important factor in the rate of increase in BLEU scores for word and morpheme-level models (see Figure 4.2). With an increase in corpus size, the OOV rate drops sharply while the translation quality improves significantly. Though OOV rates are initially low for OS and BPE-level models, they still show modest improvement in translation quality. Perhaps, additional data helps to model more lexical changes and language phenomena.
- Note that the BPE vocabulary was constant in these experiments. However, we can expect to improve the performance of BPE-level models by tuning the BPE vocabulary size for different training corpora sizes. We explore this question in the next section.

4.1.2 Effect of change in translation domain at test time

Since we are concerned with low-resource scenarios, a desirable property of subword units is robustness of the translation models to change of translation domain at test time. To study the effect of domain change, we conducted experiments on a *cross-domain translation task i.e.,* testing models trained on one domain on test data from another domain.

Pair	Char	Word	Morph	OS	BPE
pan-hin	2.08	25.17	11.81	1.85	0.14
kok-mar	10.87	36.49	9.02	3.42	0.52
mal-tam	1.14	46.69	22.44	3.46	0.08
tel-mal	0.00	40.89	11.29	2.02	0.00
hin-mal	0.00	21.64	11.28	1.65	0.00
mal-hin	0.00	46.91	23.02	2.22	0.00

TABLE 4.1: OOV rate for test set (agriculture domain) w.r.t training set (health + tourism corpus).

We trained models on a mix of tourism and health domain data (the same as described in Chapter 3). We tested these models for different translation units on an agriculture domain test set of 1000 sentences from the ILCI corpus (Jha, 2012).

4.1.2.1 OOV rate

Change of domain entails new vocabulary not observed during training. Hence, the coverage of the vocabulary in the test domain will be impacted if the model is trained on a different domain. Table 4.1 shows the OOV rates for the agriculture domain test set w.r.t to the health + tourism domain training set. We see that word and morpheme representations show a very high percentage compared to the subword-level models. In fact, there is a substantial increase in the OOV ratio compared the in-domain scenario, whereas the OOV rates for subwords are relatively stable across domains (see Table 3.9 in Chapter 3 for comparison with in-domain OOV rates). Thus, subword-level representation can mitigate data sparsity in the case of cross-domain translation.

4.1.2.2 Translation quality

Are the benefits of low OOV rate reflected in the translation quality on the cross-domain translation task? Table 4.2 shows the BLEU and LeBLEU scores for these experiments. Since the word-level model depends on coverage of the lexicon, it is highly domain dependent, whereas the subword-level models are not. Hence, translation quality is very low for the word-level models. So, even character-level models are competitive with word-level models in a cross-domain translation setting. OS and BPE-level models outperform the other translation units in a cross-domain translation setting. The LeBLEU indicates that BPE and OS-level models are better than morpheme-level models, though the BLEU scores do not show a consistent improvement.

4.2 Investigation of Design Choices and Hyperparameters

In this section, we investigate alternative design choices and hyperparameters.

Pair	Char	Word	Morph	OS	BPE
pan-hin	58.07	58.95	**59.71**	57.95	59.66†
kok-mar	17.97	18.83	18.53	**19.12**	18.42†
mal-tam	4.12	5.49	5.84	5.93	**6.75**†
tel-mal	3.11	3.26	**4.06**	3.83	3.75
hin-mal	3.85	5.18	5.99	6.24	**6.37**
mal-hin	8.42	9.92	11.12	13.36	**14.45**†

(a) BLEU scores

Pair	Char	Word	Morph	OS	BPE
pan-hin	0.866	0.825	0.868	0.863	**0.876**
kok-mar	0.647	0.641	0.643	**0.665**	0.653
mal-tam	0.301	0.261	0.378	0.452	**0.475**
tel-mal	0.246	0.198	0.238	0.297	**0.300**
hin-mal	0.281	0.336	0.354	**0.404**	0.384
mal-hin	0.439	0.371	0.466	0.548	**0.565**

(b) LeBLEU scores

TABLE 4.2: Translation quality for cross-domain translation. † indicates statistically significant difference in BLEU score between **OS** and **BPE**. BLEU score differences between (**OS**, **Word**) and (**BPE**, **Word**) are also statistically significant for all language pairs except *kok-mar* ($p < 0.05$).

4.2.1 BPE vocabulary size

With BPE, the vocabulary size is not dependent on the training corpus. Rather, it is a hyperparameter that controls the level of granularity of the vocabulary and it can be tuned to achieve the best translation quality. The vocabulary size is determined by the number of merge operations. The results for BPE-level translation discussed in the previous chapter do not explore optimal values for the number of merge operations. In this section, we discuss our experiments with varying the number of merge operations from 1000 to 4000. The translation results for these configurations are shown in Table 4.3. The optimal value of merge operations shows a modest, average increase of 1.6% and a maximum increase of 3.5% in the translation quality over the BPE-level results reported in the previous chapter across different language pairs. There is little change in the BLEU scores as the number of merge operations is varied. We also experimented with the number of merge operations greater than 4000 for some language pairs, but there seemed to be no benefit with a higher number of merge operations.

Compared to the number of merge operations reported by Sennrich et al. (2016) in a more general setting for NMT (60K), the number of merge operations is far less for translation between related languages with limited parallel corpora. We must bear in mind that their goal was different: available parallel corpus was not an issue, but they wanted to handle as large a vocabulary as possible for open-vocabulary NMT.

Lang Pair	**Number of Merge Operations**				
	Match	*1000*	*2000*	*3000*	*4000*
ben-hin	33.22	33.16	33.25	**33.30**	32.99
pan-hin	72.22	**72.28**	72.19	72.08	71.94
kok-mar	23.63	**23.84**	23.73	23.79	23.30
mal-tam	8.67	8.66	8.71	8.63	**8.74**
tel-mal	8.79	8.99	8.83	**9.12**	8.76
hin-mal	10.73	**10.96**	10.89	10.61	10.55
mal-hin	20.54	**21.23**	20.53	20.64	20.19
urd-hin	55.55	**55.69**	55.49	55.57	55.47
ben-urd	28.06	28.12	**28.19**	28.03	27.93
urd-mal	8.44	8.22	8.04	8.02	**8.57**
mal-urd	18.48	18.72	18.47	**18.79**	18.18
bul-mac	21.73	21.74	**22.27**	21.95	21.94
dan-swe	35.77	36.38	36.18	**36.61**	36.2
may-ind	59.54	**60.63**	60.24	60.35	60.15
kor-jpn	NA	10.13	9.8	**10.23**	9.92
jpn-kor	NA	**9.29**	9.23	9.02	8.96

TABLE 4.3: Translation quality for BPE-level models trained with different number of merge operations (BLEU). *Match* indicates number of merge operations chosen to match OS vocabulary size (results from previous chapter).

Yet, the low number of merge operations suggest that BPE encoding captures the core vocabulary required for translation between related languages.

4.2.2 Joint bilingual training of BPE-level models

In the BPE level translation models discussed so far, we learnt the BPE vocabulary separately for the source and target languages. In this section, we describe our experiments with jointly learning BPE vocabulary over source and target language corpora as suggested by Sennrich et al. (2016). The joint vocabulary is learnt by concatenating the source and target language corpora and then applying the BPE algorithm on the merged corpora. The idea is to learn an encoding that is consistent across source and target languages and therefore helps alignment. We expect a significant number of common BPE units between related languages.

This method will be effective only if the source and target languages use the same or compatible scripts. In the case of compatible scripts, we can represent the characters in the source and target language data in common script by one-one character mappings. In our experiments with Indic languages, we represent all the data in the Devanagari script.

Table 4.4 shows the results for joint BPE-level models. We do not see any major improvement over the monolingual BPE-level model.

Lang Pair	$\mathbf{BPE}_{best}$	$\mathbf{JB}_{1k}$	$\mathbf{JB}_{2k}$	$\mathbf{JB}_{3k}$	$\mathbf{JB}_{4k}$
ben-hin	33.30 (3k)	**33.54**	33.23	33.54	33.35
pan-hin	72.28 (1k)	**72.41**	72.35	72.13	72.04
kok-mar	23.84 (1k)	**24.01**	23.76	23.8	23.86
mal-tam	8.74 (4k)	8.6	**8.82**	8.74	8.72
tel-mal	**9.12** (3k)	8.47	8.84	8.89	8.92
hin-mal	10.96 (1k)	**11.19**	11.09	11.1	10.96
mal-hin	**21.23** (1k)	20.79	21.22	21.12	21.06
bul-mac	**22.27** (2k)	22.11	22.17	21.58	22.24
dan-swe	36.61 (3k)	36.15	**36.86**	36.51	36.71
may-ind	60.63 (1k)	**61.26**	60.98	61.11	60.66

TABLE 4.4: Translation quality for joint BPE-level models trained with different number of merge operations (BLEU).

Lang Pair	Unit	LM Order				
		5	**7**	**10**	**13**	**15**
ben-hin	OS	31.64	32.87	33.54	33.43	**33.59**
	BPE	32.79	33.10	33.22	**33.41**	33.36
mal-hin	OS	17.63	18.39	18.44	**18.56**	18.46
	BPE	20.11	20.10	**20.54**	20.46	20.48

TABLE 4.5: Effect of LM order on BLEU score for subword units (BLEU).

4.2.3 Language model order

Previous work (Vilar et al., 2007) has suggested that higher-order language models are helpful for character-level language models. In the results reported in the previous chapter, we used higher order LM models for OS and BPE-level models (10 gram) assuming that higher order LMs would be helpful for these translation units also. In this section, we perform a systematic study of the effect of the LM order on translation quality for OS and BPE-level models. For two language pair (Bengali-Hindi and Malayalam-Hindi), we tuned and tested our OS and BPE-level systems using 5, 7, 10, 13 and 15-gram language models. Table 4.5 shows the BLEU scores as the LM order varies. We see that a higher-order LM is indeed helpful for subword-level translation. Beyond 10-gram, we do see any major improvements in BLEU scores.

4.2.4 Large language models for subword-level translation

Language models are used as features in SMT models to improve the fluency of translated sentences. Typically, large monolingual corpora in addition to the target side of the parallel corpus are used for training the language model. Since word-level models have a large vocabulary, the large monolingual corpora help to reliably estimate n-gram probabilities. Given the finite size of subword-level vocabulary, is it still

Lang Pair	OS		BPE		Word	
	Small	**Large**	**Small**	**Large**	**Small**	**Large**
BLEU Score						
ben-hin	**33.54**	26.32	**33.22**	26.90	31.19	**32.47**
mal-hin	**18.54**	13.63	**20.54**	15.35	15.15	**15.18**
LM Feature Weight						
ben-hin	0.236	0.098	0.241	0.040	0.137	0.122
mal-hin	0.289	0.108	0.211	0.035	0.145	0.114

TABLE 4.6: Comparison between SMT models using large and small language models.

beneficial to use a large monolingual corpus for language modelling? We investigate this question in this section.

Experiments in the previous chapter used limited monolingual corpora for subword SMT models. In Table 4.6, we show a comparative analysis of SMT models tuned with large and small LM for word, OS and BPE-level models. For the large LM, we use a 10 million sentence monolingual corpus in addition to the target side corpus for training the Hindi language models.

We see that OS and BPE-level models show a sharp decline in the translation quality when a large LM is used. In contrast, the word-level model benefits (or at least shows no degradation) when a large language model is used. We check the weights learnt for the LM feature to understand how important the LM feature is for the model (Table 4.6). For OS and BPE-level models, the LM feature weights learnt using large LM are very small compared to the models using the small LM. In contrast, there is no significant difference between the LM feature weights for the word-level model.

Thus, a large subword-level LM is not beneficial for subword-level translation models. Subword-level translation models serve to improve learning of bilingual mappings, but subword-level n-gram language models may not necessarily help to model the target language.

4.2.5 Word *vs.*, subword-level tuning

Previous work (Nakov and Tiedemann, 2012) has suggested that tuning the system to optimize the word-level BLEU scores is better than tuning character-level BLEU scores for character-level translation models. We adopted word-level tuning for OS and BPE-level experiments described in the previous chapter. In this section, we verify if tuning the SMT system to maximize word-level BLEU scores is indeed better than subword-level tuning for OS and BPE-level models. Table 4.7 compares BLEU scores for tuning word-level and subword-level BLEU metrics. We observe that the difference in translation quality between word and subword-level tuning is statistically insignificant.

Lang Pair	OS		BPE	
	subword	*word*	*subword*	*word*
ben-hin	33.41	**33.54**	33.06	**33.22**
pan-hin	**72.46**	72.41	71.93	**72.22**†
kok-mar	23.40	**23.43**	23.48	**23.63**
mal-tam	**7.94**	7.84	8.50	**8.67**
tel-mal	8.44	**8.50**	8.75	**8.79**
hin-mal	10.30	**10.46**	10.23	**10.73**†
mal-hin	**18.66**†	18.44	20.19	**20.54**†

TABLE 4.7: Comparison between word and subword-level tuning (BLEU).

4.3 Improving Decoding Speed

The use of subword units increases the sentence length (in terms of translation units). For instance, we analysed the average length of the input sentence on four language pairs (Hindi-Malayalam, Malayalam-Hindi, Bengali-Hindi, Telugu-Malayalam) on the ILCI corpus (Jha, 2012). The average length of an input sentence for character-level representation is seven times the word-level representation, while it is four times the word-level representation for OS-level representation. The decoding time for a sentence is proportional to the length of the sentence (in terms of translation units). So, decoding time will increase for subword units. Increase in decoding time also increases tuning time since tuning involves a decoder in the loop. This makes experimentation costly and time-consuming and impedes faster feedback, which is important for machine translation research. Higher decoding time also makes deployment of SMT systems based on subword units impractical.

Exact decoding is exponential in the length of the sentence and is thus impractical (Knight, 1999). Therefore, heuristic decoding methods are popular for phrase-based SMT. One of the most popular methods is stack decoding. The time complexity of stack decoding can be expressed as shown below (Koehn, 2009, Chapter 6, p. 165):

$$O(S \times N \times T) \quad \approx \quad O(S \times N^2) \tag{4.1}$$

where,
S: stack size
N: input sentence length (in terms of basic units)
T: number of translation options for any hypothesis on the stack. T is proportional to the sentence length since any source word can be chosen for translation.

At the cost of increasing search errors, this class of decoding algorithms achieves a decoding time which is quadratic in the length of the sentence. Though stack decoding is an improvement over exact decoding, decoding is still expensive (decoding time is a quadratic function of sentence length). We study the factors affecting decoding time and suggest ways to reduce it.

4.3.1 Factors affecting decoding time

As we can see from Equation 4.1, decoding time is affected by two factors: decoder parameters (like stack size) and input sentence length.

4.3.1.1 Decoder parameters

The decoder is essentially a search algorithm, and we investigated important settings related to two search algorithms used in the *Moses* SMT system: (i) stack decoding (Jelinek, 1969; Och et al., 2001), (ii) cube-pruning (Chiang, 2007). We investigated the following parameters:

- **Stack Size** (*ss*): This parameter determines the size of the stack which maintains the best partial translation hypotheses generated at any point during stack decoding for each stack.
- **Cube Pruning Pop Limit** (*pl*): In the case of cube pruning, this parameter determines how many hypotheses should be popped for each stack.
- **Table Limit** (*tl*): Every source phrase in the phrase-table can have multiple translation options. This parameter controls how many of these options are considered during decoding.

Having a lower value for each of these parameters reduces the search space, thus reducing the decoding time. However, reducing the search space may increase search errors and decrease translation quality. For subword-level models, we hypothesize that a smaller search space can reduce the decoding time without significantly affecting translation quality since the vocabulary is significantly smaller than word-level models. We study how decoder parameters affect the decoding time-translation quality trade-off.

4.3.1.2 Input sentence length

Since the sentence is user-specified, we hardly have any control over the input length. However, recollect that we use markers to indicate word boundaries for subword-level translation. The marker convention adopted can affect the sentence length, hence we compare two popular marker conventions, which are illustrated in Table 4.8:

- **Subword Marker**: The subword at the boundary of a word is augmented with a marker character. There is one boundary subword, either the first or the last chosen as per convention. Such a representation has been used in previous work, mostly related to morpheme-level representation and in BPE-level representation for subword-level NMT (Williams et al., 2016; Sennrich et al., 2016).

Original
`this is an example of data formats for segmentation`
Subword units
`thi s i s a n e xa m p le o f da ta fo rma t s fo r se gme n ta tio n`
Subword Marker
`thi s_ i s_ a n_ e xa m p le_ o f_ da ta_ fo rma t s_ fo r_ se gme n ta tio n_`
Space Marker
`thi s _ i s _ a n _ e xa m p le _ o f _ da ta _ fo rma t s _ fo r _ se gme n ta tio n`

TABLE 4.8: Marker conventions for sentence representation with subword units (example of OS shown).

Note that Sennrich et al. (2016) mark the first subword, but this convention is equivalent to marking the last subword.

- **Space Marker**: The subword units are not altered, but the inter-word boundary is represented by a space marker. Most work on translation between related languages has used this marker convention (Nakov and Tiedemann, 2012).

For subword markers, the addition of the marker character does not change the sentence length, but can create two representations for some subwords (corresponding to internal and boundary positions). This increases the vocabulary size and may affect translation quality by introducing some data sparsity. On the other hand, space marker doubles the sentence length (in terms of words), but each subword has a unique representation.

4.3.2 Related work

It has been recognized in the past literature on translation between related languages that the increased length of subword-level translation is a challenge for training as well as decoding (Vilar et al., 2007). Subword-level alignment and decoding are computationally expensive, hence most work has concentrated on corpora with short sentences (Tiedemann, 2009; Nakov and Tiedemann, 2012; Tiedemann, 2012a). For instance, movie subtitle corpora like OPUS (Tiedemann, 2012b) have been widely used in the literature. However, this does not help process corpora with long sentences. To make subword-level alignment feasible for such corpora, Vilar et al. (2007) use the phrase-table learnt from word-level alignment as a parallel corpus for training subword-level models. The sentence/segment pairs so created are invariably shorter, but may not be complete sentences. Instead, we propose a feasible method to train SMT systems directly on long sentences at subword level. We rely on optimizing decoder parameters and using faster decoding algorithms. Related work on reducing decoding time using alternative decoder configurations and decoding algorithms (Hoang et al., 2016; Huang and Chiang, 2007; Heafield et al., 2014) have been limited to word-level models and did not necessarily involve related languages.

	Translation Quality				Relative Decoding Time			
src	**ben**	**hin**	**mal**	**tel**	**ben**	**hin**	**mal**	**tel**
tgt	**hin**	**mal**	**hin**	**mal**	**hin**	**mal**	**hin**	**mal**
Default (stack, tl = 20, ss = 100)	33.10	11.68	19.86	9.39	46.44	65.98	87.98	76.68
Stack Decoding								
tl = 10	32.84	11.24	19.21	9.47	35.49	48.37	67.32	80.51
tl = 5	32.54	11.01	18.39	9.29	15.05	21.46	30.60	41.52
ss = 50	33.10	11.69	19.89	9.36	17.33	25.81	35.76	43.45
ss = 10	33.04	11.52	19.51	9.38	4.49	7.32	10.18	11.75
+tuning	32.83	11.01	19.57	9.23	5.24	8.85	11.60	9.31
Cube Pruning								
pl = 1000	33.05	11.47	19.66	9.42	5.67	9.29	12.38	17.85
+tuning	33.12	11.30	19.77	9.35	7.68	13.06	15.18	14.56
pl = 100	32.86	10.97	18.74	9.15	2.00	4.22	5.41	5.29
pl = 10	31.93	9.42	15.26	8.50	1.51	3.64	4.57	3.84
Word-level	31.62	9.67	15.69	7.54	100.56 ms	65.12 ms	50.72 ms	42.4 ms

TABLE 4.9: Translation quality (BLEU) and relative decoding time for OS-level models using different decoding methods and parameters. Relative decoding time is indicated as a multiple of word-level decoding time. "+tuning" means that the modified parameters are used at tuning time also.

4.3.3 Experimental setup

To test the effect of decoder parameters and marker conventions, we experimented with four language pairs (Bengali-Hindi, Malayalam-Hindi, Hindi-Malayalam and Telugu-Malayalam). The chosen language pairs cover different combinations of morphological complexity between the source and target languages. We used the SMT models described in the previous chapter for decoding. Since subword-level decoding is very slow and we needed to experiment with many decoder configurations, we limit our test set to 500 sentences and experiment with orthographic syllables only. However, our approach for reducing decoding time is not specific to any subword unit. Hence, the results and observations should hold for other subword units like BPE too. The decoding was performed on a server with Intel Xeon processors (2.5 GHz) and 256 GB RAM. To compute the time to decode the test set, we use the sum of user and system time minus the time for loading the phrase-table (all reported by *Moses*).

4.3.4 Effect of decoder parameters

Table 4.9 shows the results of varying decoder parameters for OS-level models. Translation quality and decode time per sentence (in milliseconds) for word-level models are shown on the last line for comparison.

	Translation Quality				Relative Decoding Time			
src	**ben**	**hin**	**mal**	**tel**	**ben**	**hin**	**mal**	**tel**
tgt	**hin**	**mal**	**hin**	**mal**	**hin**	**mal**	**hin**	**mal**
Default (stack, tl = 20, ss = 100)	27.29	6.72	12.69	6.06	206.98	391.00	471.96	561.00
Cube Pruning pl = 1000	26.98	6.57	11.94	5.99	10.23	19.57	24.59	26.20
Word-level	31.62	9.67	15.69	7.54	100.56 ms	65.12 ms	50.72 ms	42.4 ms

TABLE 4.10: Translation quality (BLEU) and relative decoding time for character-level models using different decoding methods and parameters.

4.3.4.1 Stack decoding

With the default decoder parameters, we observed that the decoding using OS-level models is approximately 70 slower than word-level models. We can see that with a stack size of ten, the decoding is now about nine times slower than word-level decoding. This is a 7x improvement in decoding time over the default parameters, while the translation quality drops by less than 1%. When stack size = 10 is used during tuning too, the drop in translation quality is still minor (2.5% drop). On the other hand, reducing the table-limit significantly reduces the translation quality, while resulting in lesser reduction in decoding time.

4.3.4.2 Cube pruning

With a pop-limit of 1000, the decoding time is about 12 times that of word-level decoding. The drop in translation quality is about 1% with a 6x improvement over default stack decoding, even when the model is tuned with a pop-limit of 1000. Using this pop-limit during tuning also hardly impacts the translation quality. However, lower values of pop-limit reduce the translation quality.

Note: While our experiments primarily concentrated on OS as the unit of translation, we also compared the performance of stack decoding and cube pruning for character level models. The results are shown in Table 4.10. We see that character-level models are 4–5 times slower than OS-level models and hundreds of time slower than word-level models with the default stack decoding. In the case of character-level models also, the use of cube pruning (with `pop-limit` = 1000) substantially speeds up decoding (20x speedup) with only a small drop in BLEU score.

4.3.5 Effect of marker convention

For these experiments, we used the following decoder parameters: cube-pruning with cube-pruning-pop-limit = 1000 for tuning as well as testing. Table 4.11 shows the results of our experiments with different marker conventions. We see that the two conventions do not differ a lot in terms of decoding time and translation quality. Neural MT systems based on encoder-decoder architectures, particularly without

	Translation Quality				Relative Decoding Time			
src	**ben**	**hin**	**mal**	**tel**	**ben**	**hin**	**mal**	**tel**
tgt	**hin**	**mal**	**hin**	**mal**	**hin**	**mal**	**hin**	**mal**
Boundary Marker	32.83	**12.00**	**20.88**	9.02	**7.44**	11.80	17.08	18.98
Space Marker	**33.12**	11.30	19.77	**9.35**	7.68	13.06	15.18	**14.56**
Word-level	31.62	9.67	15.69	7.54	100.56 ms	65.12 ms	50.72 ms	42.4 ms

TABLE 4.11: Translation quality (BLEU) and relative decoding time for OS-level models using different marker conventions.

attention mechanism, are more sensitive to sentence length, so we presume that the boundary marker convention may be more appropriate in the scenario.

4.4 Summary

In this chapter, we conducted an empirical investigation regarding various aspects of a subword-level SMT system. The following is a summary of our investigations:

Resource Availability We observed that OS and BPE-level models clearly outperform other units in extremely data scarce scenarios. They are also better than other translation units in a cross-domain translation task showing that subword-level learning can generalize better to other domains also.

Design Choices and Hyperparameter Settings Based on our experiments, we make the following recommendations regarding subword-level translation models:

- Using a higher-order language model for subword-levels is clearly beneficial. The order of the LM needs to be tuned on a validation set.
- Use of external monolingual corpora for training subword-level LM significantly reduces translation quality. Hence, it is recommended that only the monolingual corpus from the training data be used for language modelling for the purposes of SMT.
- Only a small number of BPE operations are required for achieving good translation quality (1000–4000 merge operations). This is considerably less than the word-level vocabulary. The exact number of merge operations depends on the language and the training data size, but our experiments indicate that we need to search amongst only a small set of merge operations to determine this value.
- Learning a joint BPE vocabulary across the source and target languages does not provide any benefit over learning the vocabularies separately.

- We did not observe any significant difference between using subword or word-level tuning.

Improving Decoding Speed We found that subword-level decoding speed can be substantially improved without loss in translation quality by using cube pruning coupled with a small cube pruning pop limit. The use of these decoder parameters also reduces tuning time. A direction for future work is to investigate methods for reduction of subword alignment time. The choice of marker convention for sentence representation has a marginal impact accuracy on decoding time.

Chapter 5

Subword-level Pivot-based SMT

In Chapter 3, we showed that subword units can improve translation between related languages by utilizing morphological isomorphism and lexical similarity. In order to address translation requirements of many related languages spanning multiple domains, it must be possible to easily (or with reasonable effort) extend coverage of translation systems to new language pairs and domains. We have already shown that the subword-level SMT models trained on one domain, perform reasonably well on a new domain, thus making it possible to increase coverage on new domains. In this chapter, we address the question of increasing coverage to more related language pairs by enabling translation between related languages which do not share a parallel corpus.

5.1 Motivation

If there are n related languages, we need parallel corpora between $\frac{n(n-1)}{2}$ language pairs to cover all language pairs. This is impractical to achieve in most cases. Alternatively, we may build an *n-way* parallel corpus between all the languages — a multilingual corpus where parallel sentences across all related languages are aligned. Unless there is apriori co-ordination, it is difficult to create such an aligned corpus. Even the *Europarl* corpus, a compilation of proceedings of the European Parliament, is pairwise aligned with only English.

A reasonable scenario to assume is:

- Parallel corpora exist/can be compiled between some language pairs in the set of related languages.
- We can cover most language pairs for which no parallel corpora exists by translating through a common related language (referred to as *pivot language*). We assume the availability of source-pivot and pivot-target language parallel corpora.

A few examples to illustrate this scenario:

- Hindi serves as a *de facto* link language across India and it is possible to find parallel corpora between Hindi and other Indian languages.

- Given the influence of Tamil in southern India, it may be possible to find parallel corpora between Tamil and other major South Indian languages.
- Some minority languages may be very close to a related language that is comparatively rich with respect to availability of parallel corpora. For instance, Tulu is a language spoken by 1.7 million people in some parts of the Indian states of Kerala and Karnataka and is very close to Malayalam and Kannada, the official languages of these states. Given the close socio-cultural contacts between Tulu and Malayalam/Kannada speaking people, it is very likely that Tulu-Malalayam/Kannada parallel corpora can be mined. On the other hand, being the official languages of major Indian states, Malayalam and Kannada are likely to have parallel corpora with other related languages like Hindi and Tamil. Hence, Malayalam or Kannada can serve as pivot languages to enable translation between Tulu and other Indian languages.

If no parallel corpus is available between two languages, pivot-based SMT (Gispert and Marino, 2006; Utiyama and Isahara, 2007) provides a systematic way of using the *pivot language* to build source-target translation system. The general pivot approach makes no assumptions about relatedness between the source, pivot and target languages. In this chapter, we investigate how the relatedness between languages can be used to improve pivot-based SMT.

5.2 Pivot-based SMT for Related Languages

For translation between two related languages which do not share a parallel corpus, we propose the following pivot-based SMT approach:

> *Use a third language related to the source and target as pivot language along with subword-level representation (orthographic syllables or BPE units).*

Thus, source, pivot and target languages are all related in this scenario. We hypothesize that subword-level representation can improve pivot-based SMT because:

- The component models used to build the pivot-based model, *viz.* source-pivot and pivot-target models, are trained at the subword-level. As discussed in the previous chapters, they are better than other translation units.
- The pivoting process involves bridging the source and target languages via the shared vocabulary of the pivot language. A large vocabulary can cause data sparsity and hence hinder the ability to learn source-target translation mappings. The nature of sharing and sparsity depends on the specific pivot method used.

The subword-level pivot-based SMT system is trained in two stages.

First, we train the source-pivot and pivot-target language subword-level translation models as described previously in Chapter 3. To summarize, we build subword-level PB-SMT models with the following adaptations: (a) monotonic decoding since related languages have similar word order, (b) higher order language models (10-gram) since data sparsity is a lesser concern due to small vocabulary size (Vilar et al., 2007) and (c) word-level tuning (by post-processing the decoder output during tuning) to optimize the correct translation metric (Nakov and Tiedemann, 2012). After decoding, we regenerate words from subwords (desegmentation) by concatenating subwords between consecutive occurrences of the boundary markers.

Next, we composed the source-pivot and pivot-target models. We explore two pivoting methods for the same: triangulation and pipelining.

Triangulation (Utiyama and Isahara, 2007; Wu and Wang, 2007; Cohn and Lapata, 2007) "joins" the source-pivot and pivot-target subword-level phrase-tables on the common phrases in the pivot columns, generating the pivot model's phrase-table. It recomputes the probabilities in the new source-target phrase-table, after making a few independence assumptions, as shown below:

$$P(\bar{t}|\bar{s}) = \sum_{\bar{p}} P(\bar{t}|\bar{p})P(\bar{p}|\bar{s}) \tag{5.1}$$

where, $\bar{s}, \bar{p}$ and $\bar{t}$ are source, pivot and target phrases, respectively.

In the **pipelining** approach (Utiyama and Isahara, 2007), a source sentence is first translated into the pivot language, and the pivot language translation is further translated into the target language using the source-pivot and pivot-target translation models respectively. To reduce cascading errors due to pipelining, we consider the top-k source-pivot translations in the second stage of the pipeline (an approximation to expectation over all translation candidates). We used $k = 20$ in our experiments. The translation candidates are scored as shown below:

$$P(\mathbf{t}|\mathbf{s}) = \sum_{i=1}^{k} P(\mathbf{t}|\mathbf{p}^i)P(\mathbf{p}^i|\mathbf{s}) \tag{5.2}$$

where, s, $\mathbf{p}^i$ and t are the source, i^{th} best source-pivot translation and target sentence respectively.

5.3 Related Work

Our work straddles two strands of research in statistical machine translation: (i) subwords as basic units for translation between related languages and (ii) pivot-based machine translation.

We have discussed related work on subword-level translation between two languages in Chapter 3. In this monograph, we have proposed two units of translations for related languages (OS and BPE units). Pivot translation provides a systematic way for translation between two languages through an intermediate language. Multiple approaches to pivoting have been proposed *viz.* (i) synthetic corpus generation (Gispert and Marino, 2006), (ii) transfer-based/pipelining (Utiyama and Isahara, 2007) and (iii) phrase-table triangulation (Utiyama and Isahara, 2007; Wu and Wang, 2007; Cohn and Lapata, 2007).

Character-level units have been previously explored for pivot-based SMT involving related languages (Tiedemann, 2012a; Galušcáková and Bojar, 2012; Tiedemann and Nakov, 2013). However, they use a character-level model in just one language pair of the triple (either source-pivot or pivot-target, but not both). In these works, the pivot is related to either the source or target (but not both). In our scenario, the source, target and pivot languages are all related. Hence, subword-level translation is possible for both stages (source-pivot and pivot-target). We explore OS and BPE-level pivot-based SMT involving related languages.

5.4 Experimental Setup

Languages: We experimented with multiple languages from the two major language families of the Indian subcontinent: *Indo-Aryan* branch of the Indo-European language family (Bengali, Gujarati, Hindi, Marathi, Urdu) and *Dravidian* (Malayalam, Telugu, Tamil). We experimented with many language triples covering various combinations of the Indo-Aryan and Dravidian languages. Urdu uses an *abjad* script, while the other languages involved use Brahmi-derived *abugida* scripts.

Datasets: We used the *Indian Language Corpora Initiative (ILCI) corpus*[1] for our experiments (Jha, 2012). The data split is as follows — **training: 44,777; tuning: 1000; test: 2000** sentences. Language models for word-level systems were trained on the target side of training corpora plus monolingual corpora from various sources (same as described in Chapter 3). We used the target side of parallel corpora for morpheme, OS, BPE and character-level language models.

System details: The details of PBSMT and language training, segmentation of the corpus for different translation units is the same as described in Chapter 3. For phrase-table triangulation, we use *tmtriangulate*[2]. For linear interpolation of phrase-tables, we use *combine-ptables* (Bisazza et al., 2011), which is included in the *contrib* section of *Moses*.

[1] Available on request from `http://tdil-dc.in`

[2] `github.com/tamhd/MultiMT`

Evaluation: Just as we did in the experiments in Chapter 3, we used word-level BLEU (Papineni et al., 2002) as the primary evaluation metric and LeBLEU (Virpioja and Grönroos, 2015) as the secondary evaluation metric. We use bootstrap resampling for testing statistical significance (Koehn, 2004).

5.5 Results and Analysis

In this section, we present the results of our experiments and analyse them. We investigated the following questions:

- Are OS and BPE-level pivot models better than pivot models trained on other translation units? (Sections 5.5.1 and 5.5.2)
- Are OS and BPE-level pivot models competitive with direct translation systems? (Section 5.5.3)
- Are OS and BPE-level pivot models better than pivot models trained on other translation units for cross-domain translation tasks? (Section 5.5.4)
- Is a related pivot really necessary? How do translation models using a related pivot language compare with translation models using an unrelated pivot language? (Section 5.5.5)

5.5.1 Comparison of different subword units

We compare pivot-based SMT systems trained on different translation units: character, word, morpheme, orthographic syllable and BPE unit. We evaluate triangulation as well as pipelining approaches for pivoting.

5.5.1.1 Triangulation approach to pivot-based SMT

We first discuss results for the triangulation approach, the results being shown in Table 5.1 in terms of the BLEU score. The last row in the table summarizes the average change in BLEU scores for word, morpheme and character-level model compared to the OS and BPE-level models. The following are the major observations:

- The simplest subword unit, the character, does not show consistent improvements compared to word-level translation, in spite of a small vocabulary size. The BLEU scores are around 7% lower than word-level pivot model scores on an average. We have already discussed the shortcomings of character-level translation in Chapter 3.
- All the other subword-level pivot translation models *viz.* morpheme, OS and BPE-level, significantly outperform word-level translation. The average

Families	Lang Triple	Word	Morph	BPE	OS	Char
Indo-Aryan	mar-guj-hin	30.23	36.49	39.05	**39.81**†	34.32
	mar-hin-ben	16.63	21.04	22.46	**22.92**†	17.00
Dravidian	mal-tel-tam	4.55	6.19	**7.69**†	7.19	3.51
	tel-mal-tam	5.13	8.29	**9.84**†	8.39	4.26
Indo-Aryan +	hin-tel-mal	5.29	8.32	9.57	**9.67**	6.24
Dravidian	mal-tel-hin	10.03	13.06	**17.68**	17.26	9.12
Involving	mal-urd-hin	7.70	11.29	**16.40**	NA	7.46
Urdu	urd-hin-mal	5.58	6.64	**7.58**	NA	4.07
average % change w.r.t (BPE,OS)		*(+66,+57)%*	*(+21,+14)%*			*(+81,+66)%*

TABLE 5.1: Comparison of triangulation approach for various subwords (BLEU). Scores in **bold** indicate highest values for the language triple. † indicates the difference between OS and BPE scores is statistically significant ($p < 0.05$). NA: OS segmentations cannot be done for Urdu.

increase in BLEU score over word-level model is 37.2%, 57.7%, 63.7% respectively. Thus, word segmentation helps improve translation quality in pivot-based SMT models.

- The OS and BPE-level pivot models, which utilize lexical similarity, outperform morpheme-level pivot models. OS and BPE-level pivot models show an average increase in BLEU score of 13.9% and 20.8% respectively over morpheme-level pivot models.
- The OS and BPE-level models show improvements over word and morpheme-level models even when the languages involved belong to different language families (but share a contact relation) *viz.* Hindi-Telugu-Malayalam and Malayalam-Telugu-Hindi.
- The OS and BPE-level models are comparable in terms of BLEU score. The BPE-level pivot models show an average improvement of around 3.6% over OS-level pivot models, in terms of BLEU score. However, the improvements are not statistically significant for all language triples. The BPE units seem to be better than OS units for the morphologically richer Dravidian languages, while OS units seem to perform better for Indo-Aryan languages.
- Unlike orthographic syllabification, BPE segmentation can also be applied to languages with non-alphabetic scripts like Urdu. We see that use of BPE-level pivot-based SMT model significantly increases translation quality for translation involving such languages also. For language triples involving Urdu, we see an average increase of 103%, 74% and 30% over character, word and morpheme-level models with BPE-level models.

To summarize, we see the same trends with pivot-based translation using OS and

Families	Lang Triple	Word	Morph	BPE	OS	Char
Indo-Aryan	mar-guj-hin	0.692	0.725	0.737	**0.747**	0.713
	mar-hin-ben	0.505	0.616	0.638	**0.646**	0.577
Dravidian	mal-tel-tam	0.247	0.364	**0.426**	0.407	0.213
	tel-mal-tam	0.242	0.433	**0.485**	0.441	0.392
Indo-Aryan +	hin-tel-mal	0.291	0.376	0.420	**0.432**	0.306
Dravidian	mal-tel-hin	0.247	0.364	**0.426**	0.404	0.213
Involving	mal-urd-hin	0.328	0.436	**0.501**	NA	0.377
Urdu	urd-hin-mal	0.313	0.353	**0.420**	NA	0.323
average % change w.r.t (BPE,OS)		*(+51,+49)%*	*(+12,+8)%*			*(+42,+42)%*

TABLE 5.2: Comparison of triangulation for various subwords (LeBLEU).

BPE units as we saw in the case of direct translation in Chapter 3. The benefits of modelling lexical similarity carry over to the pivot translation scenario. Further, the improvement in translation quality is larger as compared to direct translation. Thus, OS and BPE-level pivot models are better than pivot models trained on other translation units.

5.5.1.2 Evaluation with LeBLEU metric

As discussed in the previous chapters, BLEU has many limitations when used for evaluation of translation involving morphologically rich languages. Since our experiments involve morphologically rich languages, we also evaluate our systems using the LeBLEU metric (Virpioja and Grönroos, 2015), a metric suited for morphologically rich languages. Table 5.2 shows the LeBLEU scores. We see that the LeBLEU scores show the same trends as BLEU scores *i.e.,* BPE and OS units outperform other translation units. In the remainder of this chapter, we report just BLEU scores for the sake of brevity.

5.5.1.3 Comparing triangulation and pipelining approaches to pivot-based SMT

We study if pipelining, a widely used pivoting method, also benefits from subword-level translation. Table 5.3 shows the results for the pipelining approach with different translation units. In this case too, the BPE and OS-level models are better than the word, morpheme and character-level models. Again, the BPE and OS models are comparable to each other, with differences not being statistically significant in most cases.

Further, if we compare triangulation and pipelining approaches for BPE and OS-level translation, we observe that their BLEU scores are roughly equivalent. Table 5.4 shows a comparison of translation with the two approaches. Note that this comparison cannot be done for OS-level models involving Urdu, since Urdu cannot be OS segmented. Moreover, the triangulation approach provides an additional benefit:

Lang Triple	Word	Morph	BPE	OS	Char
mar-guj-hin	34.44	37.79	**38.25**	38.11	32.62
mar-hin-ben	20.36	22.26	22.50	**22.83**	17.63
mal-tel-tam	5.73	6.62	**7.84**†	6.94	3.75
tel-mal-tam	6.71	8.10	**8.47**†	7.96	3.60
hin-tel-mal	7.22	8.51	**9.31**	**9.31**	6.03
mal-tel-hin	12.56	14.61	**17.39**	16.96	9.83
mal-urd-hin	12.20	13.52	**16.93**	NA	8.09
urd-hin-mal	7.20	7.29	**8.83**	NA	3.97
average % change w.r.t (BPE,OS)	*(27,21)%*	*(13,5)%*			*(82,63)%*

TABLE 5.3: Comparison of pipelining approach for various subwords (BLEU). † indicates the difference between OS and BPE scores is statistically significant ($p < 0.05$).

it provides an elegant framework to combine multiple pivot languages. Hence, subsequent experiments in this chapter use the triangulation approach. Please note that subsequent mentions of pivoting in this chapter refer to triangulation, unless mentioned otherwise.

The previous two sections firmly establish that OS and BPE-level pivot translation models are consistently better than models trained on other units under a range of scenarios encompassing: (i) different pivoting methods and (ii) different evaluation metrics.

5.5.2 Why is pivoting using subwords better?

Two factors can explain the improvement in pivot models due to subword-level representation.

One, the underlying source-pivot (S-P) and pivot-target (P-T) models for the OS and BPE units are better than their counterparts trained on other units. For instance, direct translation models using OS units show an average improvement of 16% and 3% in BLEU scores over the corresponding word and morpheme-level models for the language pairs in the experiments above. In Chapter 3, we have analyzed why subword-level translation models are better.

However, we observe an interesting result: the improvement in the pivot translation BLEU scores is very significant (57% and 14% improvement for OS-level models over the word and morpheme-level models). Such a significant improvement was not observed in the case of subword-level direct translation models. This improvement cannot be explained by the improvement of the underlying models alone.

Further improvement stems from the benefit subword representation provides for the triangulation process. The triangulation process involves an *inner join* on pivot language phrases common to the S-P and P-T phrase-tables. When the vocabulary is large, the number of common phrases is small causing data sparsity issues in word

Lang Triple	BPE		OS	
	pip	*tri*	*pip*	*tri*
mar-guj-hin	38.25	**39.05**†	38.11	**39.81**†
mar-hin-ben	**22.50**	22.46	22.83	**22.92**
mal-tel-tam	**7.84**	7.69	6.94	**7.19**
tel-mal-tam	8.47	**9.84**†	7.96	**8.39**†
hin-tel-mal	9.31	**9.57**	9.31	**9.67**†
mal-tel-hin	17.39	**17.68**	16.96	**17.26**
mal-urd-hin	**16.93**†	16.40	NA	NA
urd-hin-mal	**8.83**†	7.58	NA	NA

TABLE 5.4: Comparison of pipelining (*pip*) and triangulation (*tri*) approaches for OS and BPE (BLEU). † indicates the difference between *pip* and *tri* scores is statistically significant ($p < 0.05$).

Lang Triple	Word	Morph	BPE	OS	Char
mar-guj-hin	0.64	1.39	1.74	2.33	3.04
mar-hin-ben	0.58	1.36	1.71	2.6	3.47
mal-tel-tam	0.61	2.32	3.27	4.19	2.58
tel-mal-tam	0.75	2.82	4.09	2.76	2.42
hin-tel-mal	0.56	2.08	2.86	2.97	2.25
mal-tel-hin	0.55	2.28	2.85	3.56	2.57
mal-urd-hin	0.25	1.16	1.84	NA	2.05
urd-hin-mal	0.42	0.79	1.62	NA	1.47

TABLE 5.5: Ratio of triangulated to component phrase-table sizes.

and morpheme-level triangulation. On the other hand, **the OS and BPE phrase-table vocabularies are smaller, so the impact of sparsity is limited**. This effect can be observed by comparing the ratio of the triangulated phrase-table size (S-P-T) with that of the component phrase-tables (S-P and P-T). By phrase-table size, we mean the number of entries in the phrase-table. We calculate the ratio of the phrase-table size of the triangulated system (S-P-T) with the phrase-table size of the underlying phrase-tables (S-P and P-T) as shown below:

$$ratio_{\text{ptsize}} = \frac{size_{\text{S-P-T}}}{\max(size_{\text{S-P}}, size_{\text{P-T}})} \tag{5.3}$$

We observe that the size of the triangulated phrase-table is less than the size of the underlying tables for word-level, while the phrase-table size increases by a few multiples for OS and BPE-level models. These statistics are shown in Table 5.5. Thus,

Lang Triple	Word		Morph		BPE		OS	
	direct	*pivot*	*direct*	*pivot*	*direct*	*pivot*	*direct*	*pivot*
mar-guj-hin	38.87	30.23	42.81	36.49	43.19	39.05	43.69	39.81
mar-hin-ben	21.13	16.63	23.96	21.04	24.13	22.46	23.53	22.92
mal-tel-tam	6.38	4.55	7.61	6.19	8.67	7.69	7.84	7.19
tel-mal-tam	9.58	5.13	10.61	8.29	11.61	9.84	10.52	8.39
hin-tel-mal	8.55	5.29	9.23	8.32	10.73	9.57	10.46	9.67
mal-tel-hin	15.18	10.03	17.08	13.06	20.54	17.68	18.44	17.26
mal-urd-hin	15.18	7.7	17.08	11.29	20.54	16.4	18.44	NA
urd-hin-mal	6.49	5.58	7.05	6.64	8.44	7.58	NA	NA

TABLE 5.6: Pivot *vs.*, direct translation (BLEU).

subword-level translation loses fewer translation options during the triangulation process.

5.5.3 Comparison of pivot and direct translation models

We have seen that subword-level pivot translation models significantly outperform pivot models trained on other units. It is then pertinent to ask if they are competitive with *direct translation models*?[3]

To investigate this question, we compared pivot models with direct models trained on different translation units. Table 5.6 shows the results of this comparison. While the direct models are generally better than the corresponding pivot models, the following are some specific observations w.r.t translation quality for different translation units:

- The BLEU scores of morpheme and word-level pivot models are far below their corresponding direct systems (about 18% and 32% deficit, respectively). Thus pivot translation with these units is not competitive compared to direct translation.

- On the other hand, the OS and BPE-level pivot models are competitive with their best performing direct counterparts too. They achieve about 91% (OS) and 88% (BPE) of the corresponding direct model's BLEU score.

- Moreover, the OS and BPE-level pivot models outperform word-level direct models by 6% and 10% respectively, which is encouraging.

- Remarkably, the OS and BPE-level pivot models are competitive with the morpheme-level direct models. They achieve about 95% (OS) and 99% (BPE) of the morpheme-level direct system's BLEU score.

[3]The **direct translation model** refers to a translation model trained on parallel corpora between the source and target languages.

Lang Triple	Word		Morph		BPE		OS	
	direct	*pivot*	*direct*	*pivot*	*direct*	*pivot*	*direct*	*pivot*
mar-guj-hin	0.746	0.692	0.767	0.725	0.766	0.737	0.766	0.747
mar-hin-ben	0.568	0.505	0.645	0.616	0.653	0.638	0.656	0.646
mal-tel-tam	0.314	0.247	0.409	0.364	0.465	0.426	0.447	0.407
tel-mal-tam	0.41	0.242	0.511	0.433	0.530	0.485	0.534	0.441
hin-tel-mal	0.393	0.291	0.436	0.376	0.468	0.420	0.477	0.432
mal-tel-hin	0.46	0.247	0.528	0.364	0.565	0.426	0.551	0.404
mal-urd-hin	0.46	0.328	0.528	0.436	0.565	0.501	0.551	NA
urd-hin-mal	0.35	0.313	0.379	0.353	0.416	0.420	NA	NA

TABLE 5.7: Pivot *vs.,* direct translation (LeBLEU).

Lang Triple	Pivot			Direct		
	Morph	**OS**	**BPE**	**Morph**	**OS**	**BPE**
BLEU scores						
hin-tel-mal	4.72	5.96	6.00	5.99	6.26	6.37
mal-tel-hin	8.29	11.33	10.94	11.12	13.32	14.45
mal-tel-tam	4.41	5.82	5.85	5.84	5.88	6.75
LeBLEU scores						
hin-tel-mal	0.278	0.342	0.331	0.354	0.404	0.384
mal-tel-hin	0.401	0.514	0.498	0.466	0.548	0.565
mal-tel-tam	0.327	0.379	0.400	0.378	0.452	0.475

TABLE 5.8: Translation quality for cross-domain translation.

- Evaluation with LeBLEU score also indicates that the same trends hold. LeBLEU scores are shown in Table 5.7.

Note that the OS and BPE-level direct systems are the best translation systems for the source-target pair. Thus, without any parallel corpus between the source and target language, the OS and BPE-level pivot systems are competitive with the best translation systems. **These observations strongly suggest that pivoting at the subword-level can better reconstruct the direct translation system than word and morpheme-level pivot systems.**

5.5.4 Cross-domain translation

Word and morpheme-level translation models are affected by change in the domain of text being translated, since the training data may not cover some vocabulary in the new domain. On the other hand, subword-level models can be expected to be more domain-independent since a larger fraction of their vocabulary can be shared across domains and new vocabulary can be represented as subword units. We

Lang Pair	Word-level Unrelated	OS-level Related	BPE-level Related
Pivot →	*eng*	*tel*	*tel*
hin-tam	8.65	11.61	**11.91**
hin-mal	6.27	**9.67**	9.57
mal-hin	9.63	17.26	**17.68**
mal-tam	4.50	7.19	**7.69**

TABLE 5.9: Related *vs.*, unrelated pivot (BLEU).

investigated if the OS and BPE-level pivot models are indeed robust to domain change for cross-domain translation tasks. To this end, we evaluated the pivot and direct translation models trained on tourism and health domains on an agriculture domain test set of 1000 sentences from the ILCI corpus. The results are shown in Table 5.8.

For cross-domain translation too, the OS and BPE-level pivot models outperform morpheme-level pivot models and are comparable to a direct morpheme-level model. They are also competitive with the best direct models for cross-domain translation. The OS and BPE-level pivot models systems experience a much lesser drop in BLEU scores *vis-a-vis* direct models, in contrast to morpheme-level pivot models. Since morpheme-level pivot models encounter unknown vocabulary in a new domain, they are less resistant to domain change than subword-level pivot models.

5.5.5 Related *vs.*, unrelated pivot language

So far, we have considered a related pivot language with which the source and target language share parallel corpora. We also did initial studies for another likely scenario - the pivot is an unrelated language like English with which the source and target languages share parallel corpora. For this analysis, we compared the OS and BPE-level pivot models (using a related pivot language — Telugu) with the word-level translation models (using an unrelated pivot — English). Note that, OS-level units are not useful for SMT between unrelated languages. The results (Table 5.9) show that using a related pivot is substantially better than using an unrelated pivot. *Though the source and target languages are related*, the word-level pivot system can not make any use of lexical similarity. **Given that the existence of parallel corpora via an unrelated language like English is a very practical scenario, better pivot methods for utilizing lexical similarity between source and target languages when the pivot language is unrelated would be a good direction for future investigation.**

5.6 Using Multiple, Related Pivot Languages

In the previous sections, we have shown that a subword-level, pivot translation model using a related pivot language is competitive with the best direct translation model. It is natural to ask if combining multiple pivot models, each using a different pivot language, can outperform a direct translation system.

Model	mar-ben		mal-hin	
	OS	BPE	OS	BPE
best pivot	22.92 (*hin*)	22.46 (*hin*)	17.52 (*tel*)	18.47 (*guj*)
direct	23.53	24.13	18.44	20.54
all pivots	23.69	23.20†	19.12†	20.28
direct+all pivots	24.41‡	**24.49**‡	19.44‡	**20.93**‡

TABLE 5.10: Combination of multiple pivot models, corresponding to different pivot languages (BLEU). Best pivot language is mentioned in brackets. Statistically significant difference from *direct* is indicated for: *all pivots* (†) and *direct+all pivots* (‡) ($p < 0.05$).

In a general setting, the use of multiple pivot languages has shown improvement in translation quality (Wu and Wang, 2007; Dabre et al., 2015). In these cases, the pivot models were trained at the word-level. The improvement in translation quality may be attributed to increased coverage of vocabulary and linguistic phenomena due to phrase pairs contributed by each pivot model. In this section, we empirically investigate if multiple related, pivot languages are beneficial in the case of subword-level pivot models.

5.6.1 Linear interpolation of multiple pivot models

To combine multiple pivot models, we linear interpolate triangulated phrase-tables, each corresponding to a different pivot language. In our case, each triangulated phrase-table corresponding to a phrase-table. Linear interpolation of phrase-tables (Bisazza et al., 2011) assigns weights to each phrase-table and the feature values for each phrase-pair are interpolated as per these weights:

$$f^j(\bar{s}|\bar{t}) = \sum_i \alpha_i f_i^j(\bar{s}|\bar{t}) \quad (5.4)$$

$$\text{s.t} \sum_i \alpha_i = 1 \quad (5.5)$$

$$\alpha_i \geq 0 \quad (5.6)$$

where f^j is feature j, α_i is the interpolation weight for phrase-table i. Phrase-table i corresponds to the triangulated phrase-table using language i as a pivot.

5.6.2 Results and analysis

We experimented with two language pairs: Marathi-Bengali and Malayalam-Hindi. For Marathi-Bengali, we experimented with Gujarati, Hindi and Punjabi as pivot languages. For Malayalam-Hindi, we experimented with Telugu, Marathi and Gujarati as pivot languages.

Weighting	mar-ben	mal-hin
equal	23.69	**19.12**
source	23.59	19.11
target	23.67	18.96
average	**23.81**	18.98

TABLE 5.11: Comparing different strategies to assign interpolation weights for OS-level pivot models (BLEU).

Table 5.10 shows the results of combining multiple pivot models (each corresponding to a different pivot language) using linear interpolation with **equal weights** to each pivot model. We see that **the combination of multiple pivot language models outperformed the individual pivot models, and is comparable to the direct translation system**. Previous studies have shown that word and morpheme-level multiple pivot systems were not competitive with the direct system, possibly due to the effect of sparsity on triangulation (More et al., 2015; Dabre et al., 2015). Our results show that once the ill-effects of data sparsity are reduced due to the use of sub-word models, multiple pivot languages can maximize translationquality. Combining the direct system with all the pivot systems using equal-weighted interpolation further benefitted the translation system. **Thus, multilinguality through multiple pivot languages helps overcome the lack of a direct translation system between the two languages.**

Other methods to combine phrase-tables: Further improvements could possibly be achieved by tuning the interpolation weights using an held-out dataset and/or using alternative methods to combine multiple translation models like fill-up interpolation or multiple decoding paths (decoding over multiple phrase-tables). The reader can refer to Dabre et al. (2015) for an extensive empirical comparison of various methods for combining pivot models.

Assigning interpolation weights based on lexical similarity between languages: As far as this monograph is concerned, it is worth exploring one particular strategy to assign interpolation weights. Since lexical similarity is an important concern of this monograph, we investigate strategies for assigning interpolation weights based on how lexically similar the pivot language is to the source and/or target language. We assign weights proportional to lexical similarity of the pivot language to (i) source language, (ii) target language and (iii) average of (i) and (ii). Table 5.11 compares these weighting strategies for OS-level pivot models. We do not observe any substantial difference in translation quality using different weight assignment strategies based on lexical similarity compared to assigning equal weights.

Pivot		OS	BPE
pan		22.07	21.68
guj	(s)	22.54	22.37
hin	(t,a)	22.92	22.46

(a) Marathi-Bengali

Pivot		OS	BPE
tel	(s)	17.26	17.68
mar		17.52	18.46
guj	(t,a)	17.44	18.47

(b) Malayalam-Hindi

TABLE 5.12: Choice of pivot language and lexical similarity. Translation quality (BLEU) along with indicators of lexical similarity are shown. We indicate which pivot has highest LCSR similarity with the source language (s), target languages (t) as well average LCSR similarity (a).

5.7 Choice of Pivot Language and Language Relatedness

Previous studies on the choice of pivot language have been impeded by the morphological diversity of the pivot languages and morphologically poorer pivots tend to perform better (Dabre et al., 2015; Paul et al., 2013). Thus the varied levels of sparsity induced by morphological properties dictated the choice of the pivot, rather than the intrinsic properties of the pivot language. Sparsity is not a major concern with OS and BPE-level models, hence we are able to study the effect of language properties on the choice of the pivot language.

We studied if the lexical similarity of the pivot language to the source and/or target language had an impact on the translation quality. We studied Marathi-Bengali and Malayalam-Hindi translation for multiple pivot languages and used LCSR as a measure of the lexical relatedness of the two languages. Table 5.12 shows the translation quality for different pivot languages along with lexical similarity of the pivot language to the (i) source language, (ii) target language and (iii) average of (i) and (ii). We see that the choice of pivot language makes a limited but observable difference (within one BLEU point).

For Marathi-Bengali translation, Hindi is the most beneficial pivot. It exhibits highest lexical similarity with the target language as well as highest average lexical similarity to the source and target languages. For Malayalam-Hindi translation, Marathi is the most useful pivot, with Gujarati close behind. Again, we observe that Gujarati exhibits highest lexical similarity with the target language as well as equidistant from both source and target. Gujarati is lexically more similar to Marathi.

These observations suggest that a pivot language which is either closer to the target language or equidistant from both source and target is more useful than having a pivot which is closer to the source language. This provides further evidence to the observations by Paul et al. (2013) that target language features are more important for "coherent" language pairs.

5.8 Summary and Future Directions

In this chapter, we explore the use of subword-level pivot translation models for translation between related languages. The following is a summary of our major findings:

- We show that pivot translation between related languages using subword-level representation of data and a pivot language related to source and target languages is significantly better than other translation units. More significantly, subword-level pivot translation models are *competitive* with direct translation models. Subword units make pivot models competitive by (i) utilizing lexical similarity to improve the underlying S-P and P-T translation models, and (ii) reducing losses in pivoting (owing to small vocabulary).
- Through comprehensive experiments, we show that pivot translation with subword units is better than other translation units under a wide range of scenarios like:
  - genetic and/or contact relationship between languages
  - languages using different types of writing systems
  - different pivoting methods
  - cross-domain translation
  - different evaluation metrics
- We show that combining pivot models from multiple pivot languages can lead to a pivot translation model whose translation quality is *nearly equivalent* to that of the direct translation model.
- Thus, we show that the use of subwords as translation units coupled with multiple related pivot languages can compensate for the lack of a direct parallel corpus (at least in the restricted case of related languages).
- Our experiments indicate that a pivot language that is lexically closer to the target language than the source language performs better.

5.8.1 Pivot translation involving an unrelated language

For the most part, our work has focussed on pivot translation where the source, target and pivot languages are related. However, there are a couple of scenarios involving unrelated languages that are relevant:

Unrelated Pivot Language: A relevant scenario is that the source and target languages may share parallel corpora with an unrelated pivot language. For instance, English-Indian language corpora may be available. Even though the source and target languages are related, the methods discussed in this chapter cannot leverage lexical similarity while pivoting via an unrelated language. Hence, better methods to leverage source-target relatedness would be an interesting question to explore.

Pivot language related to source or target language, but not both: Translation from a set of related languages to an unrelated *lingua franca* is a major requirement *e.g.,* English to Indian languages and *vice versa.* In such cases, only some of the related languages may share parallel corpora with unrelated languages, making pivot translation necessary. Simple solutions like using a pipeline approach with subword-level translation between related languages and word-level translation between the unrelated pair provides only limited benefit. Word order divergence is a major problem for such solutions. Better solutions to handle this translation scenario need to be investigated.

5.8.2 Neural machine translation for pivot translation

In recent years, the evolving consensus in the machine translation research community is that neural machine translation performs better than traditional statistical machine translation. However, current NMT systems require large-scale resources for good performance (Koehn and Knowles, 2017). In contrast, our SMT pivot translation work is useful for low-resource settings.

Nevertheless, NMT opens opportunities for translation between related languages. NMT provides a flexible framework for integrating subword-level representation along with multilingual training (referred to as multilingual NMT). Multilingual training opens the possibility of zero-shot translation, which provides an alternative to pivoting. Currently, these ideas are being actively explored in a general setting without much regard for language relatedness (Dong et al., 2015; Firat et al., 2016a; Johnson et al., 2017). Moreover, NMT is a flexible framework which can easily incorporate unrelated languages too. In future, we plan to explore multilingual NMT in conjunction with subword representation between related languages with a focus on reducing corpus requirements.

Later in this monograph, we will discuss our work on multilingual neural transliteration between orthographically similar languages. This constitutes a step in a direction towards multilingual neural translation for related languages.

Chapter 6

A Case Study on Indic Language Translation

In the previous chapters, we have described our investigations about how language relatedness can be utilized to improve machine translation. In this chapter, we do a large-scale case study on languages of the Indian subcontinent. The relatedness between Indian languages is a major motivation for the work in this monograph, and hence it is only fair that we should extensively study these languages.

This chapter is organized as follows: Section 6.1 is a short primer about Indian languages and Section 6.2 summarizes the major similarities among Indian languages. Section 6.3 describes the datasets used in our experiments. Section 6.4 presents the results of utilizing lexical similarity for translation between Indian languages, applying the methods we discussed in Chapter 3. This includes subword-level translation as well as transliteration of untranslated words from the output of a word-level translation system. Section 6.6 discusses English-Indic language translation and shows how structural correspondence between Indian languages can be used to share source-side pre-ordering rules. Section 6.7 presents an initial study of Indian language to English translation. Section 6.8 discusses related work on large scale translation between related languages. Section 6.9 summarizes the findings of the case study, the Indian language NLP resources created as part of the case-study and points to future possible work.

6.1 A Primer on Indian Languages

This section provides a brief summary about the languages of India.

6.1.1 Language diversity

India is one of the most linguistically diverse countries of the world. According to Census of India of 2001, India has 122 major languages and 1599 other languages. These languages span four major language families. According to *Ethnologue*[1], India has a high Greenberg linguistic diversity index of 0.914 (ranked 14th in the world, the highest outside Africa and the Pacific countries of Papua New Guinea, Solomon Islands and Vanuatu).

[1] https://www.ethnologue.com/statistics/country

India is also home to some of the most widely spoken languages in the world. According to *Ethnologue*[2], seven Indian languages are amongst the top 20 spoken languages in the world: Hindi (5th), Bengali (6th), Punjabi (10th), Telugu (15th), Marathi (16th), Urdu (18th) and Tamil (20th). Around 30 languages have more than a million speakers, and these languages account for 98% of the Indian population. In addition, English is also widely spoken in India by around 125 million people, though it is not the native language of most speakers.

Given this diversity, the Constitution of India lists 22 languages in the *Eighth Schedule*[3] which have been granted a special status. India has no national language, but Hindi serves as the *official language* of the Union government for administrative purposes, with English as the *subsidiary official language*. Different states use their own official languages.

6.1.2 Language families

The following are the four major language families in India:

- **Indo-Aryan**: This is a sub-family of the larger Indo-European language family. These languages are mainly spoken in North and Central India, and the neighbouring countries of Pakistan, Nepal and Bangladesh. The nearby island countries of Sri Lanka and Maldives also speak Indo-Aryan languages (Sinhala and Dhivehi respectively). The speakers of these languages constitute around 75% of the Indian population.

- **Dravidian**: This is a language family whose speakers are predominantly found in South India, with some speakers in Sri Lanka and a few minuscule pockets of speakers in North India. Comparative linguistics has not established any conclusive links of Dravidian languages to languages outside India, so these languages could be indigenous to the subcontinent. The speakers of these languages constitute around 20% of the Indian population.

- **Austro-Asiatic**: This language family is said to be indigenous to the subcontinent and related to the Mon-Khmer languages spoken in South-East Asia. They constitute the Munda sub-family of the Austro-Asiatic family (the other being Mon-Khmer) and are primarily spoken in parts of Central India. Khasi, a non-Munda Austro-Asiatic language, is spoken in some parts North-east India. The languages of the Nicobar islands are also Austro-Asiatic. The speakers of these languages around 5% of the Indian population.

- **Tibeto-Burman**: Many languages from the Tibeto-Burman sub-family of the larger Sino-Tibetan language family are spoken in regions of India that border Tibet and South-East Asia, along the Himalayan foothills and North-East India. Most of these languages have a small number of speakers, and these are spoken in areas that are at the intersection of India, China and South-East Asia.

[2] https://www.ethnologue.com/statistics/size

[3] http://mha.nic.in/hindi/sites/upload_files/mhahindi/files/pdf/Eighth_Schedule.pdf

In addition, an endangered set of languages from the Great Andamanese language family is spoken on the Andaman islands. Moreover, no information is available on the language(s) of the Sentinelese people of the Andaman, which whom there is no external contact.

6.1.3 Writing systems

The major Indian languages have a long written tradition and use a variety of scripts. These scripts are derived from the ancient *Brahmi* script. These are *abugida* scripts where the organizing unit is the *akshar*, a consonant cluster along with an optional *matra* (vowel diacritic). All these scripts have a high grapheme to phoneme correspondence and represent almost the same set of phonemes. This similarity is useful for utilizing orthographic and phonetic similarities across these languages. However, the visual layout of the characters is very different across languages; hence, each script has its own designated range of codepoints in the Unicode standard.

However, there are many languages that do not use Brahmi derived scripts. Prominent among these is Urdu, which uses an Arabic derived script. Kashmiri, Punjabi and Sindhi use Brahmi-derived as well as Arabic-derived scripts. Many languages in Central India and North-East India, which did not have a literary tradition in the past, have adopted the Latin script or one of the various Brahmi-derived scripts in modern times.

6.2 Relatedness among Indian Languages

Underlying the vast diversity in Indian languages are many commonalities. The languages within each language family are obviously related. In addition, because of contact over thousands of years, the linguistic features of languages belonging to different language families have also undergone convergence to a large extent. Hence, linguistics typically refer to India as a *linguistic area* (Emeneau, 1956). As a consequence, they share many cognates and borrowed words — examples of these have been discussed in Chapter 1. In addition, languages of all the four major language families have the same word order: Subject-Object-Verb. In fact, the Munda languages were originally supposed to be SVO language like their South-East Asian Mon-Khmer relatives; but, they have transformed to SOV languages (Subbārāo, 2012). The only exceptions are Khasi, Nicobarese and Kashmiri (which are all SVO languages) (Abbi, 2012).

In the remainder of this section, we present a few examples of borrowing of language features across language families to illustrate the extent of convergence between Indian languages.

- **Retroflex Sounds** (Emeneau, 1956; Abbi, 2012): These sounds are present in the Indo-Aryan languages, but not found in Indo-European languages outside the subcontinent *e.g.*, the retroflex aspirated voiceless plosive [ʈʰ:] in the Hindi word ठाकुर [tʰaːkur]. They were borrowed into these languages from either the Dravidian or Austro-Asiatic languages, which possess these sounds. In general, there is a high degree of overlap between the phoneme set of Indian languages.

- **Echo Words** (Emeneau, 1956; Subbārāo, 2012): Again, this feature is unique to Indo-Aryan languages amongst the Indo-European languages. They are a standard feature of Dravidian languages. *e.g.,* चाय-वाय (chaaya-vaaya) in Hindi which typically means tea and associated snacks.

- **Dative Subjects** (Abbi, 2012): The dative subject represents a non-agentive subject, generally the experiencer. The subject is marked in the dative case, whereas the direct object is marked with the nominative case. *e.g.,* मुझे आम चाहिए (mujhe aama chahiye) which means *I want a mango.*

- **Conjunctive Participle** (Subbārāo, 2012; Abbi, 2012): These are used to conjoin two verb phrases in a manner similar to conjunction. The two verb phrases represent two sequential actions; first action expressed with a conjunctive participle. *e.g.,* वह खाना खाके जाएगा (wah khAnA khAke jAyegA) which means *He will go after eating*. The first verb (khAke, after eating) is the participle and the second one (jAyegA, go) is the main verb.

- **Quotative Particle** (Subbārāo, 2012; Abbi, 2012): It reports someone else's quoted speech. This feature is present in Dravidian, Munda, Tibeto-Burman and some Indo-Aryan languages. *e.g.,* इति (iti in Sanskrit), असं (asa in Marathi), എന്ന് (enn in Malayalam)

- **Compounds Verb** (Subbārāo, 2012; Abbi, 2012): It refers to verbs composed of a polar verb followed by a vector verb. The polar verb carries of the semantics, whereas the vector verb is limited to a finite set of words and marks certain grammatical properties of the main verb. *e.g.,* बैठ जाऊ (baiTha jaaU, sit go), हस पडा (hasa padaa, laugh fell). The word *hasa* (laugh) is the polar verb, while *padaa* (fell) is the vector verb.

- **Conjunct Verb** (Subbārāo, 2012): It refers to a noun+verb combination, where the semantics is carried by the noun while the 'light' verb/verbalizer carries various grammatical markers. *e.g.,* सलाह देना (salah dena) which means *giving advice*. The word *dena* (giving) is the verbalizer.

Given the high level of convergence, Abbi (2012) remarks that:

> **India as a linguistic area gives us robust reasons for writing a common or core grammar of many of the languages in contact.**

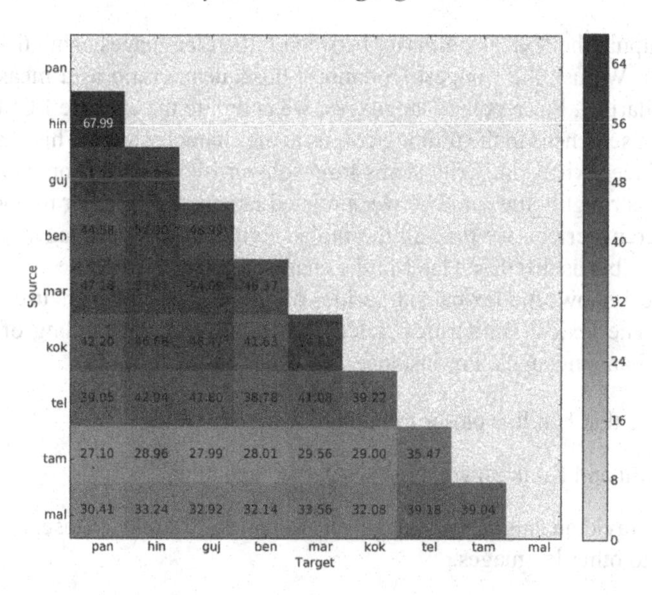

FIGURE 6.1: Lexical similarity between major Indian languages.

6.3 Dataset Used for Study

We used the Indian Language Corpora Initiative (ILCI) corpus (Jha, 2012) for the corpus. The ILCI corpus is an 11-way multilingual corpus of sentences from the health and tourism domains. The languages we experimented with are:

- **Indo-Aryan**: Hindi, Punjabi, Bengali, Gujarati, Marathi, Konkani
- **Dravidian**: Telugu, Tamil, Malayalam

These languages represent the two major language families in India. We also experiment with Indian language to English translation and *vice versa*. The data split is as follows (in number of sentences) — Train: 44,777, Tune: 1,000, Test: 2000.

6.4 Lexical Similarity between Indian Languages

As we have seen, lexical similarity is a key property between related languages and we have relied on it significantly to improve translation quality. In this section, we study the lexical similarities between Indian languages using a simple metric to provide an indicative measure of lexical similarity.

We compute the lexical similarity between Indian languages using the ILCI parallel corpus. We use the Longest Common Subsequence ratio as a measure of the lexical similarity[4]. For a pair of languages, we compute the average LCSR between every pair of sentences in the training corpus at the character level. This is an obvious approach since lexical similarity stems from subword-level correspondences and the word order is roughly the same between related languages. In order to compare text across different scripts, we map all the Indian scripts to the Devanagari script. This is easily possible as discussed later in the chapter.

Figure 6.1 shows the lexical similarities of all the language pairs used in our experiments. The lexical similarities reflect the common understanding of similarity between Indian languages. For instance,

- We see that Hindi is closer to Punjabi than to Marathi.
- Marathi and Konkani are most similar to each other.
- The Dravidian languages Malayalam, Telugu and Tamil are closer to each other, than to other languages.
- Dravidian and Indo-Aryan languages also show a reasonable level of lexical similarity between them due to contact between these language families over a long time.
- Telugu has higher similarity with Indo-Aryan languages than other Dravidian languages; Telugu speakers border Indo-Aryan speakers and hence exhibit greater lexical convergence.

Note that the lexical similarities values involving Tamil should be read with caution since the Tamil script is underspecified. The Tamil script does not have characters to represent voiced as well as aspirated plosives; these are represented by the corresponding unvoiced, unaspirated plosive.

6.5 Translation between Indian Languages

We compare translation using meaning-bearing units (words and morphemes) with translation methods which can utilize lexical similarity: (a) subword-level translation models (with OS and BPE units), (b) transliteration of untranslated words from the output of a word-level SMT system.

[4]We found a high correlation between LCSR values and manually judged lexical similarity values for a few European languages. These manual judgements were obtained from *Ethnologue* for some European language pairs, but are not available for Indian languages. Nevertheless, this result demonstrates that LCSR is a reliable metric to quantify lexical similarity automatically.

6.5.1 Word-level and morpheme-level translation

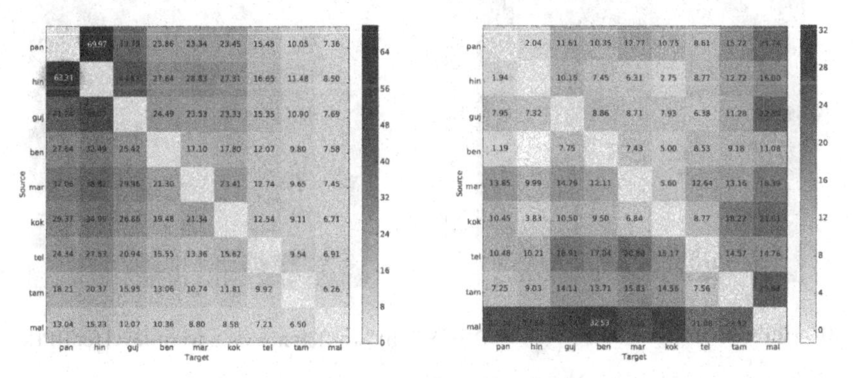

(a) Translation quality for word-level models (BLEU)

(b) % difference w.r.t. word-level for morpheme-level models

FIGURE 6.2: Word and morpheme-level translation.

The baseline word and morpheme-level models were trained as described in the previous chapters. They do not utilize the lexical similarity between languages.

Word-level translation Figure 6.2a shows translation quality for word-level translation (BLEU). We see that there is a clear partitioning of the BLEU scores as per language families. Translation between Indo-Aryan languages is the easiest, while translation between Dravidian languages is the most difficult. Moreover, translation into a Dravidian language is more difficult compared to translating from Dravidian languages. The agglutinative nature of the Dravidian languages plays an important role in making translation involving Dravidian languages challenging since it leads to data sparsity and untranslated words.

Morpheme-level translation Morpheme-level translation models are a way to address data sparsity. Figure 6.2b shows the % change in BLEU scores compared to word-level models. We observe an average increase of 12.9% in BLEU score compared to word-level models. We see a larger increase for translation involving Dravidian languages. Translation between Dravidian languages shows an improvement of 17.4%, while translation between Indo-Aryan languages shows an improvement of 7.8%.

6.5.2 Subword-level translation

While morpheme-level models reduce data sparsity and utilize morphological isomorphism between source and target languages, they cannot utilize lexical similarity between languages. To utilize lexical similarity, we train OS-level and BPE-level

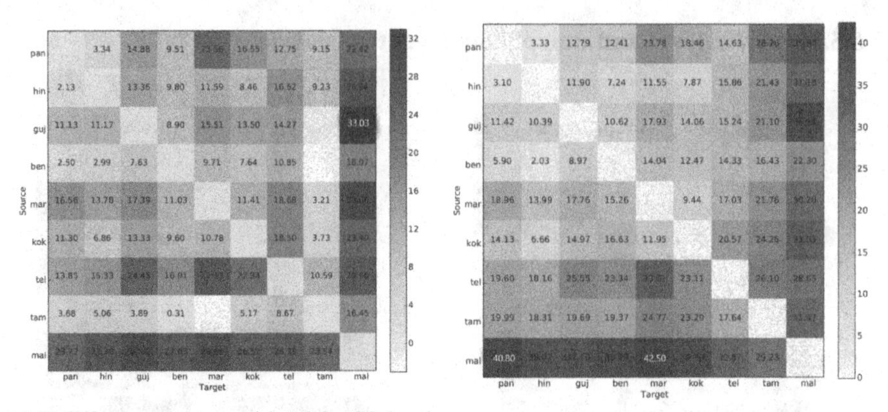

(a) % difference w.r.t. word-level for OS-level (b) % difference w.r.t. word-level for BPE-level

FIGURE 6.3: Difference in translation quality (BLEU) between subword-level models and word-level models.

models (with the number of BPE merge operations=1000), using the same procedure as described in the previous chapters.

OS-level models Figure 6.3a shows the % change in BLEU scores of OS-level models compared to word-level models. We observe an average increase of 20.2% in BLEU score compared to word-level models. Again, we see a larger increase for translation involving Dravidian languages (27.7% improvement).

BPE-level models Figure 6.3b shows the % change in BLEU scores of BPE-level models compared to word-level models. We observe an average increase of 20.2% in BLEU score compared to word-level models. Again, we see a larger increase for translation involving Dravidian languages (27.7% improvement). The BPE-level models show 6.1% improvement over morpheme-level models.

Thus, the use of lexical similarity helps improve the translation quality.

6.5.3 Discussion

6.5.3.1 Analysis by language group

While the previous sections reported differences in translation quality for every language pair, this section looks at average differences in translation quality for translation among different language groups. The two language groups under analysis are: Indo-Aryan (IA) and Dravidian (DR). Hence, the pairs of language groups for translation are: IA-IA, IA-DR, DR-IA and DR-DR *i.e.,* IA-IA indicates translation between Indo-Aryan languages and so on. Figure 6.4 shows the average % improvement in BLEU scores for translation between different language groups. ALL-ALL indicates average across all languages, irrespective of the language group. These improvements

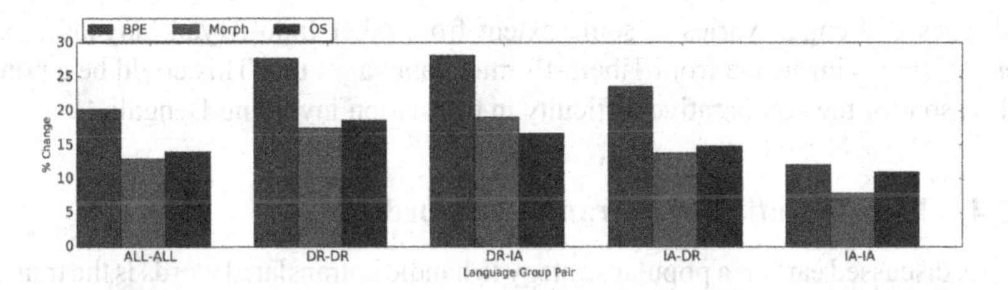

FIGURE 6.4: Comparison of average change in translation quality (BLEU) for various language groups and translation units.

Source	Ave. BLEU
hin	31.36
pan	30.16
guj	27.90
ben	20.52
mar	25.53
kok	22.97
tel	20.64
tam	16.06
mal	14.11

(a) Translating from a language

Target	Ave. BLEU
hin	39.86
pan	35.18
guj	31.33
ben	22.50
mar	22.00
kok	21.91
tel	14.97
tam	11.87
mal	9.61

(b) Translating into a language

TABLE 6.1: Average BLEU scores translating to/from different languages.

are reported for BPE, OS and morpheme-level models compared to baseline word-level models. We see that the maximum improvements are seen for Dravidian to Indo-Aryan (DR-IA) and Dravidian to Dravidian (DR-DR) translation. Thus, morphologically rich languages benefit the most from subword-level translation.

6.5.3.2 Easiest/difficult languages to translate

To derive insights into which languages are easy to translate from, we compute the average BLEU scores translating from a language into other languages (and *vice versa*). Table 6.1 shows these average BLEU scores for BPE-level models. Other translation units also show the same trends. We see that Hindi is the easiest language to translate into/from other languages. On the other hand, Malayalam is the most difficult language to translate into/from other languages. In general, translation involving Indo-Aryan languages is easier and translation involving Dravidian languages is more difficult. Morphological richness is a major distinguishing factor between Indo-Aryan and Dravidian languages and seems to be a challenging problem to address in order to further improve translation involving Dravidian languages. Marathi and Konkani, which are the most morphologically complex Indo-Aryan languages, happen to be the most difficult to translate. It is interesting to note that Bengali seems to be more difficult to translate than many other Indo-Aryan languages. The phonetics of Bengali varies to some extent from other Indo-Aryan languages, and Bengali shows influence from Tibeto-Burman languages too. This could be a potential reason for the comparative difficulty in translation involving Bengali.

6.5.4 Transliteration of untranslated words

As discussed earlier, a popular strategy to handle untranslated words is the transliteration of untranslated words in the output of an SMT system. Transliteration can achieve the purpose of translation if the source and target words are lexically similar. We experiment with two methods for transliteration of untranslated words:(a) rule-based, (b) statistical. Statistical transliteration requires a parallel transliteration corpus. We mine the parallel transliteration corpus from the parallel transliteration corpus itself. First, we summarize the transliteration mining process and the mined transliteration corpora. Subsequently, we describe the results of transliteration of the untranslated words.

6.5.4.1 Transliteration mining from parallel translation corpus

We mine transliteration pairs from the same parallel translation corpus used for training the translation system. We used the unsupervised method proposed by Sajjad et al. (2012) for mining the transliteration corpus and implemented in *Moses* (Koehn et al., 2007). Their approach models parallel translation corpus generation as a generative process comprising an interpolation of a transliteration and a non-transliteration process. The parameters of the generative process are learnt using the EM procedure, followed by extraction of transliteration pairs from the parallel corpora.

Since the transliteration pairs were mined from the translation parallel corpus, the mined pairs are representative of diverse lexical similarity phenomena — spelling variations, sound shifts, cognates and loan words. Hence, it is appropriate for training transliteration systems to transliterate untranslated words. Table 6.2 shows some examples of various kinds of mined pairs across many language pairs.

We mined a total of 1,694,576 transliteration pairs across 110 language pairs for 11 languages: Bengali, Gujarati, Hindi, Konkani, Marathi, Punjabi, Urdu, Malayalam, Tamil, Telugu and English. Table 6.3 lists the total number of transliteration pairs mined for every language pair (about 15,000 pairs on an average per pair). For some language pairs, a large number of transliteration pairs were mined as compared to others.

6.5.4.2 Rule-based transliteration

The rule-based transliteration relies on the fact that there is an almost one-one correspondence between scripts of major Indian languages, which derive from the Brahmi script. Further, the design of the Unicode standard makes rule-based transliteration very trivial. The Unicode standard assigns blocks of 128 codepoints each to various Indic scripts (*e.g.,* U+0900-U+097F for Devanagari, U+0D00-U+0D7F for Malayalam). The first 112 characters, covering the major characters, are aligned across all scripts by placing them at a common offset with respect to the start of the

Category	Example	ITRANS Transliteration
Named Entities	(అంధేరి, अंधेरी) (అక్బర్, अकबर)	(aMdherI, aMdherI) (akabara, akabara)
Spelling variations	(टेलीफोन/टेलिफोन) (बेलगाँव/बेलगाम) (फेब्रुवारी/फरवरी)	(TelIphona/Teliphona) (belagA.Nva/belagAma) (phebruvArI/pharavarI)
Tatsam words (*words borrowed as-is from Sanskrit*)	(अहंकार, അഹങ്കാരം) (करुणा, കരുണ) (चक्र, ചക്രം)	(aha.nkAra, aha NkAra.n) (karuNA, karuNa) (cakra, cakra.n)
English Loan words	(syphilis, सिफिलिस) (telephone, टेलिफोन) (counselling, কাউন্সিলিং)	(syphilis, siphilisa) (telephone, Teliphona) (counselling, kAunsili.n)
Indian origin words	(tandoori, तंदूरी) (avatar, अवतार) (yoga, योगा)	(tandoori, ta.MdUrI) (avatar, avatAra) (yoga, yogA)
Sound shifts	(केरळ, केरल)	(keraLa, kerala)
Cognates	(अंधळेपणा, अंधेपन) (कसे, कैसे) (गाढव, गधा)	(aMdhLepaNa, aMdhepan) (kase, kaise) (gaDhav, gadha)

TABLE 6.2: Examples of mined transliteration pairs.

Unicode range for that script. For instance, the Devanagari character क (*ka*, U+0915) and the corresponding Malayalam character ക (*ka*, U+0D15) are both at offset 16 within their respective Unicode ranges. This makes transliteration simply a matter of manipulating the start of the Unicode ranges.

$$char_{tgt} = \texttt{to\_char}(\texttt{to\_codept}(char_{src}) - \texttt{rstart}(lang_{src}) + \texttt{rstart}(lang_{tgt})) \quad (6.1)$$

where,
$lang_{src}$ and $lang_{tgt}$ are source and target languages respectively
$char_{src}$ and $char_{tgt}$ are source and target characters respectively
`to_codept` is a function that returns the Unicode code point (value) corresponding to the Unicode character supplied as argument
`to_char` is a function that returns the Unicode character corresponding to the Unicode code point (value) supplied as argument
`rstart` is a function that returns the codepoint (value) corresponding to the beginning of the Unicode range for the language supplied as argument.

Figure 6.5a shows the change in transliteration scores due to this simple rule-based transliteration scheme. We see a modest improvement of about 1% in BLEU score after transliteration of the untranslated words using this simple scheme. This

	hin	urd	pan	ben	guj	mar	kok	tam	tel	mal	eng
hin	-	21,185	40,456	26,880	29,554	13,694	16,608	9,410	17,607	10,519	10,518
urd	21,184	-	23,205	11,379	14,939	9,433	9,811	4,102	5,603	3,653	5,664
guj	29,550	15,019	29,434	33,166	-	39,633	35,747	12,085	22,181	11,195	6,550
mar	13,677	9,523	21,490	27,004	39,653	-	31,557	10,164	18,378	9,758	4,878
kok	16,613	9,865	21,065	26,748	35,768	31,556	-	9,849	17,599	9,287	5,560
tam	9,421	4,132	7,668	10,471	12,107	10,148	9,838	-	12,138	10,931	3,500
tel	17,649	5,680	15,598	18,375	22,227	18,382	17,409	12,146	-	12,314	4,433
mal	10,584	3,727	8,406	11,375	11,249	9,788	9,333	10,926	12,369	-	3,070
eng	10,513	5,609	8,751	7,567	6,537	4,857	5,521	3,549	4,371	3,039	-

TABLE 6.3: Statistics about mined transliteration pairs.

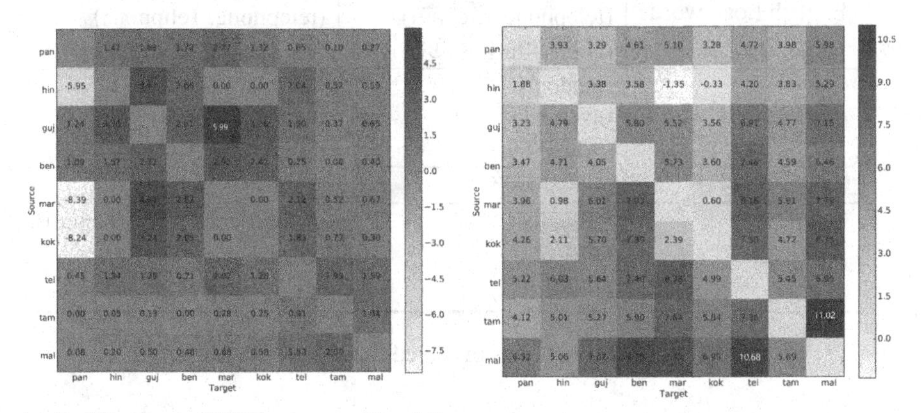

(a) % difference in BLEU w.r.t. word-level for rule-based transliteration of untranslated words

(b) % difference in BLEU w.r.t. word-level for statistical transliteration of untranslated words

FIGURE 6.5: Effect of transliteration of untranslated words from a word-level translation model.

is a very simple transliteration scheme that does not take into account phonetic variations and change in spelling conventions across the languages — it can be more appropriately referred to as *script conversion*. Hence, it is not surprising that we do not get a significant improvement in translation quality; even if the transliteration is off by a single character, it would not contribute to improvement in BLEU score. Nevertheless, it improves the perceived translation quality for users since they can read the untranslated words in the target script and guess the named entities and cognates. This is useful for *gisting*[5]/language access scenarios as envisaged by systems like *Anusaaraka* (Bharati et al., 2003). Note that no parallel transliteration corpus was required for transliteration.

[5]`http://daily.unitedlanguagegroup.com/stories/editorials/machine-translation-gisting`

The simple script conversion mechanism has a few advantages:

- Just a single rule (Equation 6.1) is sufficient to achieve transliteration/script conversion among many Indian languages. This is another illustration of the principle of utilizing language relatedness for reusing resources across languages. In this case, the orthographic similarity between languages was utilized in the design of the Unicode standard, which in turn enabled sharing of script conversion rules.
- A common transliteration scheme can be defined to represent Indian scripts using the Latin script by utilizing the correspondences in the character sets. The ISO 15919[6] and the OM (Ganapathiraju et al., 2005) transliteration schemes precisely do this. While the ISO 15919 scheme required non-ASCII characters, the OM scheme requires only ASCII characters found on a conventional keyboard making its usage convenient. ITRANS[7] is a popular transliteration scheme for Devanagari script which uses only ASCII characters. We extend ITRANS to support other Indian scripts, and provide an implementation for ITRANS to Indian script conversion and vice-versa[8] (based on a previous implementation for Devanagari[9]). All transliterations for the Indian language examples in this monograph have been generated using this implementation.

6.5.4.3 Statistical transliteration

We use a statistical transliteration system based on phrase-based SMT system techniques for transliterating untranslated words. Figure 6.5b shows the improvement in translation quality due to statistical transliteration of untranslated words. We observe an average improvement of 5.3% in translation quality (BLEU). Note that this improvement is less than the improvement achieved using subword-level translation. This provides further evidence that subword-level translation is better than the transliteration of untranslated words in the output of an SMT system.

6.6 Translation from English to Indian Languages

English is a Subject-Verb-Object language, while the canonical word order in all major Indian language is Subject-Object-Verb (though these languages are free word order to a reasonable extent). This is the fundamental divergence that needs to be bridged for translation from English to Indian languages. We first built a baseline PBSMT system, followed by two source-side pre-reordering systems.

[6] https://en.wikipedia.org/wiki/ISO_15919
[7] https://www.aczoom.com/itrans
[8] https://github.com/anoopkunchukuttan/indic_nlp_library
[9] http://www.alanlittle.org/projects/transliterator/transliterator.html

6.6.1 Baseline PBSMT

We trained word-level, phrase-based SMT systems for translation from English to Indian languages using *Moses* (Koehn et al., 2007), with the *grow-diag-final-and* heuristic for symmetrization of word alignments and the *msd-bidirectional-fe* model for lexicalized reordering (Tillmann, 2004). We tuned the trained models using Batch MIRA (Cherry and Foster, 2012) with default parameters. We trained 5-gram language models on the target side corpus with the Kneser-Ney smoothing using SRILM (Stolcke et al., 2002).

6.6.2 Reusing source-side pre-ordering rules

Phrase-based SMT system is weak at modelling long-distance word-order divergences, and lexicalized reordering is not sufficient to address word order divergences between English and Indian languages. Hence, we experimented with a source-side pre-ordering system. The source-side pre-ordering system pre-processes the source sentence, reordering it to conform to the word order of the target language at train, tune and test time. This provides two advantages:

- The typical stack decoder used for PBSMT cannot look at long distance reorderings due to computational considerations, resulting in the poor performance of PBSMT for translation between structurally divergent language pairs. With source-side pre-reordering, the source and target languages almost have the same word order; hence, exploring a small reordering window during decoding is sufficient.
- Longer phrases can be extracted by the phrase extraction algorithm resulting in more fluent output.

Ramanathan et al. (2008) postulated the following generic transformation principle going from English to Hindi word order:

$$SS_m VV_m OO_m C_m \quad \leftrightarrow \quad C'_m S'_m S' V'_m V' O'_m O' \tag{6.2}$$

where,
S: Subject, O: Object, V: Verb, C_m: Clause modifier
X': Corresponding constituent in Hindi.
X is S, O or V
X_m: modifier of X
The following is an example of the application of the generic rule:

They showed that source-side pre-ordering rules, based on the above principle, improve English-Hindi translation quality. This principle holds across all Indian languages, hence we hypothesize that the same rules will benefit translation from English to any Indian language. We call this the *generic pre-ordering* system for English-Indian language translation.

English	$\overbrace{\text{the hero}}^{S}\ \overbrace{\text{of the movie}}^{S_m}\ \overbrace{\text{shot}}^{V}\ \overbrace{\text{the scene}}^{O}\ \overbrace{\text{quickly}}^{V_m}$
English Pre-ordered	$\overbrace{\text{the movie of}}^{S_m}\ \overbrace{\text{the hero}}^{S}\ \overbrace{\text{quickly}}^{V_m}\ \overbrace{\text{the scene}}^{O}\ \overbrace{\text{shot}}^{V}$
Hindi	फिल्म के (S_m) नायक ने (S) जल्दी (V_m) दृश्य (O) शूट किया (V)
Hindi (ITRANS)	$\overbrace{\text{philma ke}}^{S_m}\ \overbrace{\text{nAyaka ne}}^{S}\ \overbrace{\text{jaldI}}^{V_m}\ \overbrace{\text{dRîshya}}^{O}\ \overbrace{\text{shUTa kiyA}}^{V}$

Pre-ordering	pan	hin	guj	ben	mar	kok	tel	tam	mal
None	15.83	21.98	15.80	12.95	10.59	11.07	7.70	6.53	3.91
Generic	17.06	23.70	16.49	13.61	11.05	11.76	7.84	6.82	**4.05**
Hindi-tuned	**17.96**	**24.45**	**17.38**	**13.99**	**11.77**	**12.37**	**8.16**	**7.08**	4.02

TABLE 6.4: English to Indian language translation (BLEU).

Further, Patel et al. (2013) refined Ramanathan et al. (2008)'s rules based on an error analysis of English-Hindi translation output. We postulate that the rules resulting from analysis of the English-Hindi system would benefit translations to other Indian languages too. We call this refined system the *Hindi-tuned pre-ordering* system. We applied these rules that had previously been tested with Hindi alone to all Indian languages in our experiments. We wanted to validate if the same rules can be successfully reused for other Indian languages, taking advantage of the structural correspondence between Indian languages.

6.6.3 Results and discussion

Table 6.4 shows BLEU scores for various English-Indian language translation systems. The following are the major observations:

- We observe that both that the source-side pre-ordering systems improve translation quality for all the Indian languages, though they had previously been developed and tested for English-Hindi translation. The *generic* system shows a 5% average improvement in BLEU score, whereas the *Hindi-tuned* system shows a 9% improvement in BLEU score.

- The average improvement over all Indian languages roughly corresponds to the improvement in English-Hindi translation, showing that the pre-ordering rules are just as useful for other Indian languages as they are for Hindi.

- The *Hindi-tuned* system, which was customized based on English-Hindi error analysis, is better than the *generic* system for other Indian languages. On an average, the *Hindi-tuned* system improves the BLEU scores by 4% over the *generic* system.

	pan	hin	guj	ben	mar	kok	tel	tam	mal
PBSMT	22.12	24.70	19.21	17.39	16.48	16.27	13.60	11.78	8.87

TABLE 6.5: Indian language to English translation (BLEU).

- Both the reordering systems show better improvement for Indo-Aryan languages compared to Dravidian languages. Since there are some syntactic divergences between Indo-Aryan and Dravidian languages, there may be a case for customizing the rules for Dravidian languages.

6.7 Translation from Indian Languages to English

We trained phrase-based SMT systems for translation from Indian languages to English using the *Moses* (Koehn et al., 2007), with the *grow-diag-final-and* heuristic for symmetrization of word alignments and the *msd-bidirectional-fe* model for lexicalized reordering (Tillmann, 2004). We tuned the trained models using Batch MIRA (Cherry and Foster, 2012) with default parameters. We trained 5-gram language models on the target side corpus with the Kneser-Ney smoothing using SRILM (Stolcke et al., 2002).

Table 6.5 shows the BLEU scores. Obviously, lexicalized reordering is not sufficient for handling structural divergences between Indian languages and English. These results are initial results, which can be improved with source-side pre-reordering and syntax-based MT methods. Currently, these approaches are not feasible since most Indian languages do not have a constituency or dependency parser available. In the spirit of the utilizing language relatedness, parsers for Indian languages must define the same set of dependency relations, and a common parsing framework for Indian languages must be defined. Efforts in this direction are underway by different research groups: (a) dependency annotation scheme for Indian languages based on traditional Indian Paninian grammar has been defined (Begum et al., 2008), (b) dependency annotated corpora are being created and (c) parsers which work across Indian languages are being experimented with (Bharati et al., 2009; Bhat, 2017).

6.8 Related Work

Our work is most similar to Koehn (2005) and Koehn et al. (2009), who describe and analyze translation systems for all language pairs in the *Europarl* and *Acquis*

Communautaire corpus respectively. The former work spans 11 languages and builds 110 translation systems. The latter work spans 22 languages builds 462 translation systems. The languages covered are all European languages. They build word-level PBSMT systems for these languages pairs. Both the works report that language relatedness affects translation quality, but do not propose any solution to utilize language relatedness. Both the works also suggest that rich morphology is a challenge to building machine translation systems.

Tyers and Alperen (2010) have created the *South-East European Times Parallel Corpus* of Balkan languages and trained word-level PBSMT systems for 72 language pairs. While the authors envision this as a step towards pan-Balkan translation, the work does not study the effect of language relatedness or its utilization for improvement of translation. On a smaller scale, Thu et al. (2016) have developed the Asian Language Treebank containing small parallel corpora for many South-East Asian languages. Recently, Ding et al. (2016) have reported baseline PBSMT system scores for two language pairs from this corpus.

English to multiple Indian language phrase-based SMT systems have been explored by Post et al. (2012) (six Indian languages from crowd-generated corpora) and the *Anuvadaksh* project (eight Indian languages)[10]. No evaluation of the latter system is available in the public domain, except for the English-Hindi SMT engine (Ramanathan et al., 2008; Patel et al., 2013). To the best of our knowledge, ours is the first large-scale study of Indian language to Indian language SMT systems, and Indian language to English SMT systems. Subsequent to our work, Agrawal (2017) have compared NMT and SMT systems for 110 Indian language pairs. They have focussed on improving the NMT systems using linguistic features and monolingual corpora, but have not investigated how language relatedness can be used to improve NMT systems. They found that NMT systems made fewer morphological and syntax/agreement errors, but more lexical choice errors compared to SMT systems.

Most of the work in pan-Indian language MT has involved rule-based MT systems. The *AnglaBharati* system (Sinha et al., 1995) is an *English-to-Indian* language based pseudo interlingua-based MT system which harnesses the common characteristics of Indian languages in the syntax transfer stage. The syntax transfer change is common to all languages and generates a pseudo-target language output corresponding to Indian language word order. The *Sampark* system[11] (Anthes, 2010) is a transfer-based system for translation between nine Indian language pairs that uses a common lexical transfer engine, whereas minimum structural transfer is required between Indian languages. The emphasis is on detailed morphological analysis to enable accurate lexical transfer and target generation. Lexical similarity has not be harnessed in any significant measure in these systems.

[10]`http://tdil-dc.in/index.php?option=com_vertical&parentid=72`
[11]`http://sampark.iiit.ac.in`

6.9 Summary and Future Work

6.9.1 Findings

In this chapter, we have presented an extensive case-study on translation involving nine major Indian languages covering 72 language pairs. We experiment with the different translation units discussed in previous chapters. The following are the major findings:

1. The case study provides further evidence that OS and BPE-level translation models perform significantly better than word and morpheme-level models. We also observe that OS and BPE-level translation models are better than approaches relying on the transliteration of untranslated words in the output a word-level translation model.

2. The major trends we observed in our previous experiments also hold: (a) subword-level translation models are more beneficial for morphologically rich Dravidian languages, (b) they are also effective when only contact relation exists between the languages.

3. Based on translation quality, we see clear partitioning of translation pairs by language family. For instance, translations involving Indo-Aryan languages can be done with a high level of accuracy, whereas those involving Dravidian languages are extremely difficult. This suggests that SMT approaches customized to language family pairs must be investigated.

4. Rich morphology of Indian languages, especially Dravidian languages, is a major factor impacting translation quality. For instance, it is easiest to translate to/from Hindi (a language with a relatively isolating morphology. On the other hand, translation involving Malayalam (a highly agglutinative language) is the most difficult.

We also report English to Indian language as well as Indian language to English translation results. We observe that word order divergence between English and Indian languages impacts translation quality. We show that English-Indian language translation quality can be improved by source-side pre-ordering. We show that the same set of pre-ordering rules gives improvements in translation quality across all Indian languages, showing that resources developed for one language can be reused for other related languages.

The case-study thus spans 90 language pairs (90 Indian-Indic, nine English-Indic and nine Indic-English language pairs). To the best of our knowledge, this is: (a) the first large-scale study specifically devoted to utilizing language relatedness to improve translation between related languages, (b) the largest study of translation involving Indian languages.

6.9.2 Future work

The following are some tasks that can be done in future to extend the case-study and, in general, improve machine translation for Indian languages:

- Subject to availability of parallel corpora, we would like to extend this study to more Indian languages. Some of the major Indian languages that have not been included in this study are Kannada, Urdu, Assamese. In particular, we would like to: (a) cover all 22 languages listed in Eighth Schedule of the Constitution of India, (b) cover major languages from the Austro-Asiatic family (*e.g.*, Santali, Mundari, Khasi) and the Tibeto-Burman family (*e.g.*, Bodo, Meitei).
- It would be interesting to have an exhaustive study of pivot-based SMT for Indian languages, experimenting with different pivot languages as well as combinations of pivot languages. These studies could be useful for further insights into the choice of pivot languages and relatedness among the Indian languages.
- Investigation of a common framework for addressing structural divergence in Indian language to English translation is an important research direction to pursue.
- An extensive case-study for Neural Machine Translation involving Indian languages is another relevant direction of work. In particular, it would be interesting to explore multilingual models trained at the subword-level, where lexical similarity and pooling of parallel corpus resources can be tested together.

Part II

Machine Transliteration

Chapter 7

Utilizing Orthographic Similarity for Unsupervised Transliteration

In the previous chapters, we have addressed machine translation between related languages. We utilized lexical similarity, morphological isomorphism and structural correspondence to improve translation quality and reduce data and linguistic resources. This chapter begins the second part of the monograph addressing transliteration involving related languages. In the context of transliteration, we are interested in utilizing orthographic similarity between related languages.

Supervised transliteration is a simpler task than translation since the vocabulary is smaller and decoding is monotonic. Utilizing language relatedness between the source and target languages is probably not required for bilingual, supervised transliteration problems. Hence, we focus on more challenging transliteration tasks where language relatedness is crucial to improving transliteration accuracy: unsupervised transliteration and multi-lingual transliteration (joint training of multiple languages pairs). Moreover, an investigation with transliteration serves as a good starting point to address unsupervised and multilingual translation between related languages.

In this chapter, we discuss our proposed method for unsupervised transliteration between related languages which utilizes orthographic similarity. We define the unsupervised transliteration task as follows:

> Learn a transliteration model ($\mathcal{TX}$) from the source language (F) to the target (E) language given their respective monolingual word lists, W_F and W_E respectively.

7.1 Motivation

Why do we need unsupervised transliteration? The best performing transliteration solutions are supervised, discriminative learning methods which learn transliteration models from parallel transliteration corpora (Bisani and Ney, 2008; Jiampojamarn et al., 2008). However, such corpora are available only for some language pairs. It is also expensive and time-consuming to build a parallel transliteration corpus.

This limitation can be addressed in three ways:

1. Train a transliteration model on mined parallel transliterations. The transliterations can be mined from monolingual comparable corpora (Jagarlamudi and

Daumé III, 2012) or parallel translation corpora (Sajjad et al., 2012). However, it may not be possible to mine enough transliteration pairs to train a system for most languages (Irvine et al., 2010).

2. Transliterate via a bridge language (Khapra et al., 2010) when transliteration corpora involving bridge languages are available.

3. Learn transliteration models in an unsupervised setting using only monolingual word lists.

We explore unsupervised transliteration, addressing shortcomings in previous work on this task (Ravi and Knight, 2009; Chinnakotla et al., 2010). These approaches have two major limitations in existing unsupervised transliteration approaches:

1. Lack of linguistic signals to direct the learning process

2. Limited use of context since their models are character-based

Due to this knowledge-lite approach, these models perform poorly. To mitigate this knowledge deficit, we incorporate prior knowledge in the unsupervised transliteration model by utilizing orthographic similarity between the source and target languages. This helps to learn a character-level transliteration model. Future, we incorporate contextual information in the transliteration model by learning a substring-based model.

7.2 Related Work

Unsupervised transliteration has not been widely explored. Chinnakotla et al. (2010) generate transliteration candidates using manually developed character mapping rules and rerank them with a character language model. The major limitations are: (i) character transliteration probability is not learnt, so there is undue reliance on the language model to handle ambiguity and (ii) significant manual effort for good coverage of mapping rules.

Ravi and Knight (2009) propose a decipherment framework based approach (Knight et al., 2006) to learn phoneme mappings for transliteration without parallel data. In theory, it should be able to learn transliteration probabilities and is a generalization of Chinnakotla et al. (2010)'s approach. But its performance is very poor due to lack of linguistic knowledge and has a reasonable performance only when a unigram *word-level LM* is used, instead of the conventional n-gram character-level LM used in transliteration models. The *word-level LM* signal essentially reduces the approach to a lookup for the generated transliterations in a target language word list; the method resembles transliteration mining. It will perform well only if the unigram LM has a good coverage of all named entities in the source word list. It may be difficult to find the exact surface words in the unigram LM for morphologically rich target languages since named entities can inflect for case, number, gender, *etc.*

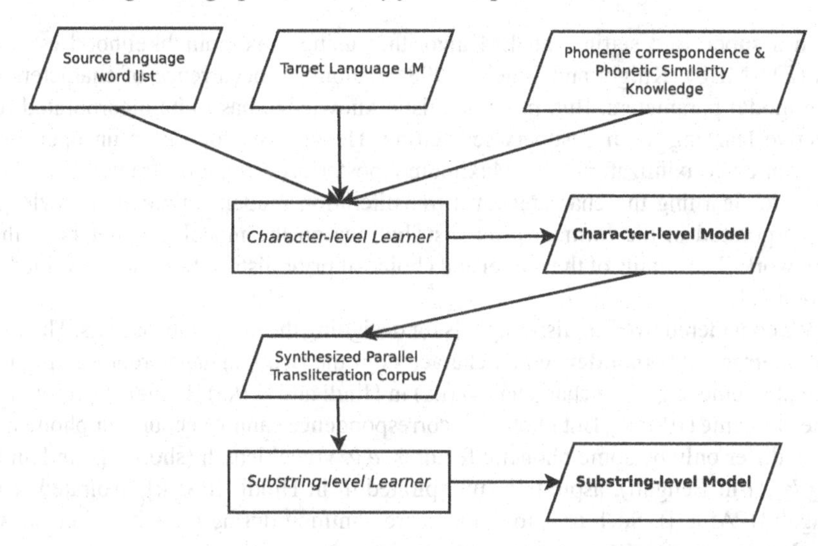

FIGURE 7.1: Overview of proposed approach.

Our character-level model approach is a further generalization of Ravi and Knight (2009)'s work since it also allows modelling of prior linguistic knowledge in the learning process. This overcomes the most significant gap in their work.

7.3 Unsupervised Substring-based Transliteration

In this section, we give a high-level overview of our approach for learning a substring-based transliteration model in an unsupervised setting (depicted in Figure 7.1). The inputs are monolingual lists of words, W_F and W_E, for the source (F) and target (E) languages respectively. Note that these are neither parallel nor comparable lists. We need a **phonemic representation** of the words *i.e.,* a sequence of phonemes representing the word.

For Indic scripts, which are used in our experiments, we use the orthographic representation itself as the phonemic representation since there is high grapheme to phoneme correspondence. Hence, we use the terms *character* and *phoneme* interchangeably.

The training is a two-stage process as described below. First, character mappings are learnt followed by learning of substring mappings by bootstrapping the character-based model. This process is analogous to phrase-based statistical machine translation, where phrase pairs are extracted from word aligned sentence pairs.

7.3.1 Character-level learning

In the first stage, character transliteration probabilities are learnt from the monolingual word lists.

In a supervised setting, an EM algorithm using maximum likelihood estimation (EM-MLE) (Knight and Graehl, 1998) exploits co-occurrence of characters to learn model parameters. But, prior linguistic knowledge has to be incorporated for effective learning in an unsupervised setting. Hence, we propose an unsupervised Expectation-Maximization with Maximum Aposteriori estimation framework (EM-MAP) for learning the character-level transliteration model. Linguistic knowledge is incorporated in the form of prior distributions on the model parameters in this framework. The details of the model and choice of prior distributions are described in Section 7.4.

We considered two linguistic signals for designing the prior distributions. The first is **phonemic correspondence** *i.e.*, characters in the two languages representing the same phoneme. *e.g.*, The characters क (ka) in Hindi and ক (ka) Bengali represent the same phoneme (IPA: k). But phonemic correspondence cannot account for phonemes which differ only by some phonetic features. *e.g.* vowel length (short इ [i] in Hindi, long ঈ [I] in Bengali), aspiration (unaspirated क in Hindi [IPA: k], aspirated খ in Bengali [IPA: k^h]). Such transformations are common during transliteration, so we use **phonetic similarity** as our second linguistic signal.

7.3.2 Substring-level learning

In the second stage, we learn a discriminative, log-linear model with **arbitrary substrings as the unit of transliteration**. For learning the substring based model, a **pseudo-parallel** transliteration corpus is first synthesized using the character-level model. We discuss Stage 2 in detail in Section 7.5.

We illustrate the need for substring-level models with an example. In Indic scripts, the *anusvaara* (nasalization diacritic) can map to any of the five nasal consonants depending on the consonant following the *ansuvaara* in the source word. So, in the Hindi-Bengali pair (चंबल [*ca.mbala*], চম্বল [*cambala*]), the *anusvaara* (.m) maps to the nasal consonant (m) since the next character is the labial consonant *ba*. This shows the need for contextual information to resolve transliteration ambiguities. Substring-based models, which learn substring mappings like *.mba* → *mba*, are one way to incorporate contextual information and have been shown to perform better in a supervised setting (Sherif and Kondrak, 2007). Contextual information is especially important in an unsupervised setting.

7.4 Character-based Unsupervised Transliteration

In the first stage, we **learn character transliteration probabilities** from monolingual word lists. The generative story for the training data (the source language corpus, W_F) is explained below using a noisy channel model.

An unknown target language word $\mathbf{e}$ is selected one character at a time using a character language model $P(\mathbf{e})$ for the target language. The observed source language word $\mathbf{f}$ is generated from the target word by a channel whose properties are

Algorithm 2 Train character-level model

1: **procedure** UNSUP-CHAR(W_F, LM_E)
 ▷ LM_E: char-level language model for E
2: $\Theta \leftarrow$ initialize-params()
3: $i \leftarrow 0$
4: **while** $i \leq N$ **do** ▷ N: Number of iterations
5: $W_{E'} \leftarrow$ c-decode(W_F, Θ, LM_E)
6: $A \leftarrow$gen-alignments($W_F, W_{E'}$)
7: $\Theta \leftarrow \arg\max_{\Theta} \mathcal{Q}'_{W_F}(\Theta)$ ▷ m-step
8: **if** converged(Θ) **then**
9: **break**
10: $\sigma \leftarrow$ e-step(A, Θ)
11: **return** Θ

represented by the transliteration probability distribution (Θ). In our model, each character e in the target word $\mathbf{e}$ generates either 1 or 2 characters of the source word $\mathbf{f}$.

The probability of an observed example $\mathbf{f}$ is $P(\mathbf{f})$. We want to account for the target word generating $\mathbf{f}$ in the data likelihood equation. *However, the target language word* $\mathbf{e}$ *is a latent variable in the unsupervised setting*, so we need to compute the expectation over all possible values of $\mathbf{e}$.

$$P(\mathbf{f}) = \sum_{\mathbf{e}} P(\mathbf{f}|\mathbf{e})P(\mathbf{e}) \tag{7.1}$$

We use Knight and Graehl (1998)'s transliteration model where, the word pair $(\mathbf{f}, \mathbf{e})$ is generated by successively selecting one or more source characters f for each target character e as per a latent alignment a with probability $P(f|e) = \theta_{f,e}$. Following Ravi and Knight (2009), we restrict our model to 1-1 and 1-2 character mappings from target to source characters.

$$P(\mathbf{f}|\mathbf{e}) = \sum_{\mathbf{a}} \prod_{i=1}^{|\mathbf{e}|} \theta_{f_{a_i}, e_i} \tag{7.2}$$

The **likelihood of the observed dataset** is given by:

$$L(\Theta) = \prod_{\mathbf{f} \in W_F} \sum_{\mathbf{e}} P(\mathbf{e}) \sum_{\mathbf{a}} \prod_{i=1}^{|\mathbf{e}|} \theta_{f_{a_i}, e_i} \tag{7.3}$$

In the supervised framework of Knight and Graehl (1998), Maximum Likelihood Estimation of parameters is performed using the EM algorithm, where the alignment structure a is the latent variable. Similarly, we can also define an MLE objective for the unsupervised task, and attempt to solve it via EM. The M-step objective over the entire training set (W_F) would be:

$$\mathcal{Q}_{W_F}(\Theta) = \sum_{\mathbf{f} \in W_F} \left\{ \sum_{\mathbf{e}} \left\{ \delta_{\mathbf{e},\mathbf{f}} \sum_{\mathbf{a}} \left[\sigma_{\mathbf{a},\mathbf{f},\mathbf{e}} \sum_{f,e} n_{f,e,\mathbf{a}} \log \theta_{f,e} \right] + \log P(\mathbf{e}) \right\} \right\} \quad (7.4)$$

s.t

$$\forall e \in C_E, \sum_{j=1}^{j=|C_F|} \theta_{f_j,e} = 1$$

where, $\delta_{\mathbf{e},\mathbf{f}} = P(\mathbf{e}|\mathbf{f})$, $\sigma_{\mathbf{a},\mathbf{f},\mathbf{e}} = P(\mathbf{a}|\mathbf{e},\mathbf{f})$ are conditional probabilities of the latent variables computed in the E-step. These are computed using the previous iteration's parameter values, whose values are fixed in the current iteration. $n_{f,e,\mathbf{a}}$ is the number of times characters e and f are aligned in the alignment structure $\mathbf{a}$. C_F and C_E are the character sets of the source and target languages respectively.

This unsupervised objective differs from the supervised learning objective in this respect: the target word $\mathbf{e}$ is also a latent variable. Hence, the expectation over all latent target language words $\mathbf{e}$ occurs only in the unsupervised objective.

In the supervised scenario, the EM algorithm is able to discover hidden alignments and estimate transliteration probabilities based on co-occurrence of characters. In the absence of parallel transliteration corpora, co-occurrence is no longer a learning signal and it is not possible to learn the character transliteration probabilities reliably. To compensate for this, we define **Dirichlet priors** (D_e) over each character transliteration probability distributions (Θ_e), which can be used to encode linguistic knowledge. We use Dirichlet priors since the character transliteration probabilities P(f|e) are modelled using a multinomial distribution. Hence, we have:

$$P(\theta_{f_1,e}...\theta_{f_{|C_F|},e}) = D_e(\alpha_{f_1,e}...\alpha_{f_{|C_F|},e}) \quad (7.5)$$

Hence, the prior log-probability for the parameters is given by:

$$\mathcal{Q}_{prior}(\Theta) = \sum_{e \in C_E} \log D_e(\alpha_{f_1,e}...\alpha_{f_{|C_F|},e}) \quad (7.6)$$

By incorporating the prior log-probability into the MLE objective for the M-step, we obtain the following **EM-MAP training objective** for the M-step over the entire training set (W_F).

$$\mathcal{Q}'_{W_F}(\Theta) = \mathcal{Q}_{W_F}(\Theta) + \mathcal{Q}_{prior}(\Theta) \quad (7.7)$$

s.t

$$\forall e \in C_E, \sum_{j=1}^{j=|C_F|} \theta_{f_j,e} = 1$$

where, C_F and C_E are the character sets of the source and target languages respectively.

In the unsupervised setting, the target word (e) is also a latent variable. As seen in Equation 7.4, the M-step requires computing an expectation over all latent variables (target word and alignments). Given the target word, it is possible to enumerate all alignments of the word pairs, but it is not possible to enumerate all possible strings (**e**). *Hence, we approximate the expectation over **e** by a **max** operation* (the so-called Viterbi approximation).

The modified objective ($\mathcal{Q}'_{W_F}$) for the M-step effectively means: With the current set of parameter values, we decode the source words to generate the target words, creating a synthetic parallel transliteration corpus. Then, the M-step updates can be done using Knight and Graehl (1998)'s supervised framework. The resulting update equation for the transliteration probabilities in the **M-step** is:

$$\theta_{f,e} = \frac{1}{\lambda_e}\left\{\alpha_{f,e} - 1 + \sum_{\mathbf{f} \in W_F} \sum_{\mathbf{a}} \sigma_{\mathbf{a},\mathbf{f},\mathbf{e}} n_{f,e,\mathbf{a}}\right\} \quad (7.8)$$

where, λ_e is a normalizing factor and **e** is the best transliteration of **f** as per the previous iteration's parameters.

The **E-step** update to compute the conditional probabilities of the latent alignment variables is given by:

$$\sigma_{\mathbf{a},\mathbf{f},\mathbf{e}} = \frac{1}{Z} \times \prod_{i=1}^{|\mathbf{e}|} \theta_{f_{a_i},e_i} \quad (7.9)$$

where, Z is a normalizing factor.

The training procedure can thus be understood to follow a *decode-train-iterate* paradigm. Algorithm 2 shows the procedure for character-level training. In each iteration, a pseudo-parallel corpus by decoding W_F using the current set of parameters (Line 5, Viterbi approximation) from which updated parameters are learnt (Line 7) and alignment probabilities recomputed (Line 10).

7.4.1 Linguistically informed priors

In this section, the different prior distributions we designed to encode phonetic knowledge about characters is described. These are instantiations of Dirichlet priors (D_e), which serve as a conjugate prior to the multinomial distribution ($\theta_{f,e}$). The hyperparameters ($\alpha_{f,e}$) of D_e determine the nature of the prior distribution. They can be interpreted as additional, virtual alignment counts of the character pair (f, e) for maximum likelihood estimation of $\theta_{f,e}$.

Phoneme Correspondence (PC) Prior: This simply establishes a one-one correspondence (denoted by $\widehat{=}$) between the same phonemes (or characters representing the same phonemes). It does not capture the notion of similarity between characters.

$$\alpha_{f,e} = \beta \quad \text{if} f \widehat{=} e \tag{7.10}$$
$$= 0.01 \quad \text{elsewhere} \tag{7.11}$$

Phonemic correspondence is also used to initialize the transliteration probabilities for the EM algorithm (*PC Init*):

$$\theta_{f,e}^{\text{init}} = \frac{\alpha_{f,e}}{\sum\limits_{x \in C_F} \alpha_{x,e}} \tag{7.12}$$

Phonetic Similarity Priors: This prior captures similarity between two phonemes based on their phonetic properties. The phonetic properties of a phoneme can be encoded as a bit vector (v) as explained in Section 7.4.2. We experimented with two priors based on phonetic properties.

- Cosine Prior: It is based on the cosine similarity between the two bitvectors.

$$\alpha_{f,e} = \gamma_c \cos(v_f, v_e) \tag{7.13}$$

- Sim1 Prior: Cosine similarity tends to produce very diffused transliteration probability distributions. We propose a modified prior (called *sim1*) which tries to alleviate this problem by making the phonemic differences sharper.

$$\alpha_{f,e} = \gamma_s \frac{5^{v_f \cdot v_e}}{\sum\limits_{x \in C_F} 5^{v_x \cdot v_e}} \tag{7.14}$$

β, γ_c and γ_s are scale factors for the Dirichlet distribution.

7.4.2 Extracting phonetic features for Indic scripts

We now describe a method for deriving phonetic correspondences and constructing phonetic features vectors for Indic scripts. Indic scripts generally have a one-one correspondence from characters to phonemes in the scripts. Hence, each character is represented by a feature vector representing its phonetic properties as described in Table 7.1. The feature vector is represented as a bit vector with a bit for each value of every property.

The logical character set is roughly the same across all Indic scripts, though the visual glyph varies to a great extent. So phonemic correspondence can be easily determined for Unicode text since the first 85 characters of all Indic scripts are aligned to each other by virtue of having the same offset from the start of the script's codepage. These cover all commonly used characters. There are a few exceptions to this simple mapping scheme, most of which can be handled using simple rules.

Feature	Possible Values
Basic Character Type	vowel, consonant, anusvaara, nukta, halanta, others
Vowel Features	
Length	short, long
Strength	weak, medium, strong
Status	Independent, Dependent
Horizontal position	Front, Back
Vertical position	Close, Close-Mid, Open-Mid, Open
Lip roundedness	Close, Open
Consonant Features	
Place of Articulation	velar, palatal, retroflex, dental, labial
Manner of Articulation	plosive, fricative, flap, approximant (central or lateral)
Aspiration	True, False
Voicing	True, False
Nasalization	True, False

TABLE 7.1: Phonetic features for Indic scripts.

Notable among these is the Tamil script, which does not have characters for aspirated as well as voiced plosives, so the corresponding unvoiced, unaspirated plosive characters are used to represent these sounds too. In the phonetic feature representation of such characters for Tamil, both the voiced as well as unvoiced bits and aspirated/unaspirated bits are set on, reflecting the ambiguity in the grapheme-to-phoneme mapping.

7.5 Bootstrapping Substring-based Models

In the second stage, we train a **discriminative, log-linear transliteration model which learns substring mappings**. We use the log-linear model proposed by Och and Ney (2002) for statistical machine translation and analogous transliteration features. The features are: substring transliteration probabilities, weighted average character transliteration probabilities and character language model score. The conditional probability of the target word **e** given the source word **f** is:

$$P(\mathbf{e}|\mathbf{f}) = \prod_{i=1}^{NP} P(\bar{e}_i|\bar{f}_i) = \prod_{i=1}^{NP} \exp \sum_{j=0}^{NF} \lambda_j g_j(\bar{f}_i, \bar{e}_i) \quad (7.15)$$

where, $\bar{f}_i$ and $\bar{e}_i$ are source and target substrings respectively, λ_j and g_j are feature weight and feature function respectively for feature j, NP number of substrings and NF is number of features.

We synthesize a **pseudo-parallel transliteration corpus** ($W_F, W_{E'}$) for training the discriminative model by transliterating the source language words (W_F) using the character-level model from the first stage. Since the top-1 transliteration may be incorrect, we consider the top-k transliterations to improve the odds that the pseudo-parallel corpus contains the correct transliteration. For **training**, the pseudo-parallel corpus contains k transliteration pairs for every source language word. For **tuning** the feature weights, we use a small held-out set of top-1 transliteration pairs from the pseudo-parallel corpus, since it likely to be the most accurate one.

We run **multiple iterations** of the discriminative training process, with each being trained on the pseudo-parallel corpus synthesized using the previous iteration's models. The models in subsequent iterations are **bootstrapped** from the earlier models. The training continues for a fixed number of iterations although other convergence methods can also be explored. Like the MAP-EM solution for the first stage, the second stage also uses a *decode — train — iterate* paradigm for learning a substring-based model.

7.6 Experimental Setup

7.6.1 Data

We experimented on the following Indian language pairs representing two language families: Bengali→Hindi, Kannada→Hindi, Hindi→Kannada and Tamil→Kannada. Bengali (ben) and Hindi (hin) are Indo-Aryan languages, while Kannada (kan) and Tamil (tam) are Dravidian languages. We used 10K source language names as training corpus, which were collected from various sources.

We evaluated our systems on the NEWS 2015 Indic dataset. We created this set from the English to Indian language training corpora of the NEWS 2015 shared task (Banchs et al., 2015) by mining name pairs which have English names in common. 1500 words were selected at random to create the test set. The remaining pairs are used to train and tune a supervised transliteration system for comparison. The training sets are small, the number of name pairs being: 2556 (ben-hin), 4022 (kan-hin), 3586 (hin-kan) and 3230 (tam-kan).

7.6.2 System details

We trained the character-level unsupervised transliteration systems with source language word lists using a custom implementation [1]. We set the value of the scaling factors ($\beta, \gamma_c, \gamma_s$) to 100. Viterbi decoding was done with a bigram character language model, followed by re-ranking with a 5-gram character language model.

We trained the substring-level discriminative transliteration models as well as a supervised transliteration system using the *Moses* (Koehn et al., 2007) machine

[1] `https://github.com/anoopkunchukuttan/transliterator`

Method	ben-hin			hin-kan			kan-hin			tam-kan		
	A_1	F_1	A_{10}	A_1	F_1	A_{10}	A_1	F_1	A_{10}	A_1	F_1	A_{10}
PC_Init	12.72	68.95	18.94	0.00	44.76	0.07	0.20	48.84	0.54	0.00	44.46	0.27
Rule	16.13	74.60	16.13	**13.75**	**79.67**	13.75	12.90	79.29	12.90	10.25	68.49	10.25
Initialization: PC_Init+												
PC_Prior	**18.27**	**75.50**	27.04	12.53	77.32	17.89	**27.69**	**81.06**	**43.55**	13.49	**69.85**	**29.06**
Cosine Prior	17.74	75.09	26.57	11.38	75.08	18.09	17.54	77.69	32.86	13.21	69.44	26.64
Sim1 Prior	18.07	75.25	**29.05**	11.72	75.61	**20.26**	19.69	78.18	37.84	**13.55**	69.74	28.19
Supervised	32.06	83.03	63.32	30.01	85.93	69.37	54.23	90.05	80.04	30.74	81.62	64.33

TABLE 7.2: Results for character-based model (% scores).

translation system with default parameters. Batch MIRA (Cherry and Foster, 2012) was used to tune the Stage 2 systems with 1000 name pairs and supervised systems with 500 name pairs. The tuning set for the Stage 2 systems were drawn from the top-1 transliterations in the synthesized, pseudo-parallel corpus; no true parallel corpus is used. Monotone decoding was performed. We used a 5-gram character language model trained with Witten-Bell smoothing on 40K names for all target languages. We ran Stage 2 for five iterations.

For a rule-based baseline, we used the script conversion method implemented in the *Indic NLP Library*[2] (Kunchukuttan et al., 2015) which is based on phonemic correspondences.

7.6.3 Evaluation

We used top-1 accuracy based on exact match (A_1) and Mean F-score (F_1) at the character-level as defined in the NEWS shared tasks as our evaluation metrics (Banchs et al., 2015). We also used top-10 accuracy as an evaluation metric (A_{10}), since applications like MT and IR can further disambiguate with context information available to these applications.

7.7 Results and Discussion

Table 7.2 shows the results for the rule-based system and various character-based unsupervised models. Table 7.3 shows results for substring-level models bootstrapped from different character-based models. Results of supervised transliteration on a small training set are also shown in both tables.

7.7.1 Baseline models

Parameter initialization with phoneme correspondence mappings and add-one smoothing prior (*PC_Init*) is comparable to Ravi and Knight (2009)'s method and performs very poorly as reported in their work too. We also experimented with

[2] `http://anoopkunchukuttan.github.io/indic_nlp_library`

re-ranking the results using a unigram word based LM — our approximation to Ravi and Knight (2009)'s use of a word based LM — and its accuracy is comparable to *PC_Init*. The unigram LM was trained on a corpus of 185 million and 42 million tokens for Hindi and Kannada respectively. Thus, this knowledge-lite approach cannot learn a transliteration model effectively.

Rule-based transliteration (*Rule*) performs significantly better than *PC_Init*. The phonetic nature of Indic scripts makes the rule-based system a stronger baseline, yet this simple approach does not ensure high accuracy transliteration. Phonetic changes like changes in manner/place of articulation, voicing, *etc.* make transliteration non-trivial and phonetic correspondence is not sufficient to ensure good transliteration.

7.7.2 Effect of linguistic priors

The addition of linguistically motivated priors (*PC_Prior, Cosine_Prior, Sim1_Prior*) significantly improves the transliteration accuracy over the *PC_Init* approach. There is a significant improvement in top-1 accuracy over the *Rule* approach too for two language pairs (12%, 31% and 133% increase for Bengali-Hindi, Tamil-Kannada and Kannada-Hindi pairs respectively), but a drop of 9% for the Hindi-Kannada pair. A major reason for lower accuracy of Hindi-Kannada pair is an important difference between Kannada and Hindi writing conventions. Unlike Hindi, Kannada assumes an implicit *schwa* vowel sound at the end of a word unless another vowel or nasal character terminates the word. Therefore, the vowel suppressor character (*halanta*) must be generated during Hindi-Kannada transliteration. Our method is poor at this generation, but conversely, it does better at deletion of *halanta* for Kannada-Hindi transliteration. There is a substantial improvement for the Tamil-Kannada pair also, even though there are some grapheme-to-phoneme ambiguities in the Tamil script.

In general, the phonemic correspondence prior results in better top-1 accuracy, whereas priors using phonetic similarity give better top-10 accuracy. The phonetic similarity based priors are smoother compared to the sparse *PC_Prior* since they capture character similarity. This allows them to discover more character mappings, resulting in better top-10 accuracy at the cost of a drop in top-1 accuracy. The sparse *sim1* prior outperforms the smoother cosine similarity prior.

7.7.3 Effect of learning substring mappings

Substring-based transliteration improves the top-1 as well as top-10 accuracy significantly over the underlying character-based models. Across languages, the best substring-based models improve top-1 accuracy by up to 11% and top-10 accuracy by up to 25% over the best character-based models. Therefore, it is clear that contextual information can be harnessed in an unsupervised setting to substantially improve transliteration accuracy.

The iterative procedure is beneficial and transliteration accuracy increases as improved models are built in successive iterations. Large gains are particularly observed in top-10 accuracy.

Stage 1 Model	Iterations	ben-hin A_1	ben-hin A_{10}	hin-kan A_1	hin-kan A_{10}	kan-hin A_1	kan-hin A_{10}	tam-kan A_1	tam-kan A_{10}
PC_Init	1	14.12	22.89	0.00	0.06	0.53	4.50	0.07	1.35
	5	15.26	25.43	0.00	0.47	1.01	8.53	0.40	2.76
PC_Init+PC_Prior	1	18.74	29.38	13.21	19.31	**28.43**	44.96	15.31	32.23
	5	19.34	32.73	13.21	20.39	21.10	45.03	**19.29**	37.15
PC_Init+Cosine prior	1	18.86	29.65	12.40	20.94	17.94	39.92	15.71	32.91
	5	18.94	31.33	12.94	23.92	16.73	44.76	18.88	36.08
PC_Init+Sim1 prior	1	19.28	34.34	12.6	23.85	19.62	47.98	16.99	34.86
	5	**20.54**	**37.61**	**13.82**	**25.88**	18.55	**50.27**	18.95	**38.50**
Supervised		32.06	63.32	30.01	69.37	54.23	80.04	30.74	64.33
Mined Pairs		26.97	51.34	-	-	-	-	-	-
Bridge Languages		25.97	58.23	18.22	52.85	33.60	67.88	13.01	42.28

TABLE 7.3: Results for substring-based model (% scores).

While *PC_Prior* gives best top-1 accuracy results at the character-level, substring-based models bootstrapped from the *Sim1* prior give better results for both top-1 and top-10 accuracy metrics. Since the *Sim1* prior based character model has better top-10 accuracy, the pseudo-parallel corpus created using this mode is likely to be better than one created using *PC_Prior*. We also observe that substring-based models built without using phonetic priors cannot improve over much over the baseline transliteration.

7.7.4 Illustrative examples

- Phonetic similarity based priors were able to discover mappings between similar phonemes. *e.g.* The Bengali word ওয়েস্ট (keolAdeo) was correctly transliterated to the Hindi word केवलादेव (kevalAdeva), due to discovery of similarity between the labial sounds (v) and (o).
- Substring-based models made use of the context to learn the correct transliteration. *e.g.* The Bengali word কেওলাদেও (oyaesTa) was correctly transliterated to the Hindi word वेस्ट (vesta), since the model learnt the substring mapping *oya→ve*.
- Substring-based models could make the right choice between short and long vowels (a major source of errors in character-based models).

7.7.5 Comparison with other approaches

We compared our best substring-based model (based on *sim1* prior) with a supervised system and the following resource-constrained transliteration systems built using:

- **Mined pairs** from a translation corpus: We experimented with Bengali-Hindi on the *Brahminet* mined pairs corpus (Kunchukuttan et al., 2015). Mined corpora involving Kannada were not available.

- **Bridge languages:** We used the NEWS 2015 corpus for our experiments with English as the bridge language.

The results of these experiments are shown in Table 7.3. The accuracies of the substring-based system are less than the accuracies of the other methods. This is not unexpected since these methods use more parallel resources than the substring-based approach. But, note that the top-10 accuracy of the substring-based model is comparable to the top-1 accuracies of other approaches. Hence, the substring-based model may be sufficient for an application, like MT or cross-lingual IR, which uses the output of a transliteration system. In MT, untranslated words are replaced by their top-k transliterations. A language model can choose among the transliterations based on context while re-ranking resulting candidate sentences.

7.7.6 Applicability to different languages and scripts

Our experiments span four widely-spoken languages from the two major language families in the Indian subcontinent (Indo-Aryan [IA] and Dravidian [DR]). All languages use different Brahmi-descended scripts. We show improvements in IA→IA, DR→DR, DR→IA and IA→DR transliteration.

The first stage leverages the phonetic nature of Indic scripts to obtain phonetic representations from orthographic representations. There are at least 19 languages in India where this condition holds, so at least 171 language pairs can use this approach without grapheme to phoneme converters. Many other scripts/language pairs also show high grapheme-phoneme correspondence[3]. Our method, though, can also be applied to non-phonetic scripts also by representing the training data at the phonemic level using grapheme-to-phoneme converters.

The second stage makes no assumptions about language or script.

7.8 Summary and Future Work

In this chapter, we addressed the problem of unsupervised transliteration between related languages. We proposed a two-stage, iterative bootstrapping approach to utilize the orthographic similarity between languages for improving transliteration. Orthographic similarity helps incorporate rich phonetic as well as contextual knowledge. Phonetic knowledge is incorporated through a novel design of prior distributions for character-based learning, while context is incorporated via substring-based learning.

The following are our findings and contributions:

- Our approach vastly outperforms Ravi and Knight (2009)'s unsupervised transliteration method and outperforms a rule-based baseline by up to 50% for top-1 accuracy on multiple language pairs.

[3] `http://www.omniglot.com` lists grapheme-phoneme correspondences for many language/script combinations.

- We show that substring-based models are superior to character-based models. Ours is the first approach to use a substring-based model for unsupervised transliteration. In the transliteration mining literature, substring based models have been used for distinguishing transliterations from non-transliterations (Klementiev and Roth, 2006; Chang et al., 2009) or filtering out *bad* (low-scoring) transliteration pairs found in the discovery process Sajjad et al. (2011).

- The top-10 accuracy of our systems is comparable with the top-1 accuracy of supervised systems, but *requires only monolingual resources*: word lists for Indic languages using Brahmi-derived scripts or phoneme dictionaries for other languages.

- Ours is the first work to use phonetic feature vectors for transliteration. Previously, phonetic feature vectors have been used for transliteration mining. as opposed to transliteration mining. Tao et al. (2006) show improvement in transliteration mining performance using phonetic feature vectors resembling the ones we have used. Jagarlamudi and Daumé III (2012) use phonemic representation based interlingual projection for multilingual transliteration mining.

As future work, we would like to explore design of better phonemic correspondence and phonetic similarity functions, incorporation of phonetic knowledge while learning substring mappings (Stage 2) and better methods of handling noisy transliterations in the bootstrapping process. Finally, we would like to explore unsupervised translation between related languages. The proposed iterative, bootstrapping framework may be a good starting point for utilizing lexical similarity between related languages for translation.

Chapter 8

Multilingual Neural Transliteration

The transliteration problem has been extensively studied in literature for a variety of language pairs (Karimi et al., 2011). Previous work has looked at the most natural scenario — training on a single language pair. In the previous chapter, we addressed the scenario when no parallel transliteration corpus exists between two orthographically similar languages *i.e.,* unsupervised transliteration. In this chapter, we address another scenario: we have at our disposal parallel transliteration corpora from multiple language pairs. We explore joint training of transliteration models for multiple language pairs. We refer to this task as *multilingual transliteration.*

8.1 Motivation

Multi-task learning involves training multiple tasks in parallel while exploiting commonalities across these tasks. This helps improve the performance of each individual task by sharing information across multiple tasks using a shared representation. One way of sharing information could be via sharing parameters in a neural network among multiple tasks. For instance, various NLP tasks like POS tagging, NER, Chunking and semantic role labelling try to analyze various aspects of languages, hence we could train them jointly (Collobert et al., 2011).

Multilingual transliteration can be seen as an instance of *multi-task learning,* where training each language pair constitutes a task. Multi-task learning works best when the tasks are related to each other, so sharing of information across tasks is beneficial. Thus, multilingual transliteration can be beneficial, if the languages involved are related. Since orthographically similar languages share writing systems and phonetic properties, the transliteration tasks are clearly related. We can utilize this relatedness by sharing the vocabulary across all related languages. The grapheme-to-grapheme correspondences enable vocabulary sharing. It helps transfer knowledge across languages while training. For instance, if the network learns that the English character *l* maps to the Hindi character ल (*la*), it would also learn that *l* maps to the corresponding Kannada character ಲ (la). Data from both Kannada and Hindi datasets will reinforce the evidence for this mapping. A similar argument can be made when both the source and target languages are related. The grapheme-grapheme correspondences arise from the underlying phoneme-phoneme correspondences, and these are established due to consistent grapheme-phoneme mappings. Hence, we believe that transliteration involving orthographically similar languages can benefit from multilingual training.

Due to the utilization of language relatedness, the benefits that are typically ascribed to multi-task learning (Caruana, 1997) may also apply to multilingual transliteration:

- Since related languages share characters, it is possible to share representations across languages. This may help to generalize transliteration models since joint training provides an inductive bias which prefers models that are better at transliterating multiple language pairs.
- The training may also benefit from implicit data augmentation since training data from multiple language pairs is available. From the perspective of a single language pair, data from other language pairs can be seen as additional (noisy) training data. This is particularly beneficial in low-resource scenarios.
- Multi-task learning can provide a regularization effect, by avoiding complex models that overfit for a single language pair. Joint training may prefer simpler explanations that explain all the language pairs.

Finally, multilingual training opens the possibility of *zeroshot transliteration i.e.*, transliteration between language pairs which have not been seen during training.

8.2 Related Work

General Transliteration Methods Previous work on transliteration has focussed on the scenario of bilingual training. Till recently, the best-performing solutions were discriminative statistical transliteration methods based on phrase-based statistical machine translation (Bisani and Ney, 2008; Jiampojamarn et al., 2008, 2009; Finch and Sumita, 2010). Recent work has explored bilingual neural transliteration using the standard neural encoder-decoder architecture (with attention mechanism) (Bahdanau et al., 2015) or its adaptions (Finch et al., 2015, 2016). Finch et al. (2016)'s model, using target bidirectional LSTM with model ensembling, has outperformed the state-of-the-art phrase-based systems on the NEWS shared task datasets. On the other hand, we focus on multilingual transliteration with the encoder-decoder architecture or its adaptations. To the best of our knowledge, the task of multilingual transliteration has not been addressed in the past. The two strands of work are obviously complimentary, and better methods for bilingual transliteration can improve multilingual transliteration also.

Multilinguality and Transliteration To the best of our knowledge, ours is the first work on multilingual transliteration. Jagarlamudi and Daumé III (2012) have proposed a method for *transliteration mining* (given a name and candidate transliterations, identify the correct transliteration) across multiple languages using grapheme to IPA mappings. Note that their model cannot generate transliterations; it can only rank candidates. This is done by using grapheme to IPA mappings to learn a common interlingual space to be able to map names across languages into a common vector space. Some literature mentions multilingual transliteration (Surana and Singh, 2008; He et al., 2017; Prakash, 2012; Pouliquen et al., 2005) or transliteration mining (Klementiev and Roth, 2006; Yoon et al., 2007), but in these cases *multilingual* refer to methods which work with multiple languages (as opposed to joint training — the sense of the word *multilingual* as we use it).

Multilingual Translation Our work on multilingual transliteration is motivated by recently proposed multilingual neural machine translation architectures (Firat et al., 2016a). Broadly, these proposals can be categorized into two groups:

- One group consists of architectures that specialize parts of the network for particular languages: specialized encoders (Zoph and Knight, 2016), decoders (Dong et al., 2015) or both (Firat et al., 2016a).
- The other group tries to learn more compact networks with little specialization across languages by using a joint vocabulary (Johnson et al., 2017; Lee et al., 2017).

For multilingual transliteration, we adopt an approach that is closer to the latter group since the languages under consideration use compatible scripts resulting in a shared vocabulary. We specialize just the output layer for target languages, while sharing the encoder, decoder and character embeddings across languages. In this respect, we differ from Johnson et al. (2017)'s model. They share all network components across languages, but add an artificial token at the beginning of the input sequence to indicate the target language.

Multilinguality applied to NLP Tasks Our work adds to the increasing body of work investigating multilingual training for various NLP tasks like POS tagging (Gillick et al., 2016), NER (Yang et al., 2016; Rudramurthy et al., 2016) and machine translation (Dong et al., 2015; Firat et al., 2016a; Lee et al., 2017; Zoph and Knight, 2016; Johnson et al., 2017) with a view to learn models that generalize across languages and make effective use of scarce training data.

Zeroshot Transliteration We use the multilingual models to address zeroshot transliteration. *Zeroshot* transliteration using bridge/pivot language has been explored for statistical machine transliteration (Khapra et al., 2010) as well as neural machine transliteration (Saha et al., 2016). Unlike previous approaches which pipelines bilingual transliteration models, we propose zeroshot transliteration that pipelines multilingual transliteration models. We also propose a pivot zeroshot transliteration method, a scenario which has been explored for machine translation by (Johnson et al., 2017), but not investigated previously for transliteration. In our zeroshot model, sequences from multiple source languages are mapped to a common encoder representation without the need for a parallel corpus between the source languages. Another work, the correlational encoder-decoder architecture (Saha et al., 2016), maps source and pivot languages to a common space, but requires a source-pivot parallel transliteration corpus.

8.3 Multilingual Transliteration Learning

We first formalize the multilingual transliteration task and then describe our proposed solution.

8.3.1 Task definition

The multilingual transliteration task involves learning transliteration models for l language pairs $(s_i, t_i) \in \mathbf{L}$ $(i = 1 \text{ to } l)$, where $\mathbf{L} \subset S \times T$, and S, T are sets of source and target languages respectively. The languages in each set are orthographically similar. S and T need not be mutually exclusive.

We are provided with parallel transliteration corpora for these l language pairs $(D_i,\ \forall i = 1 \text{ to } l)$. The goal is to learn a joint transliteration model for all language pairs which minimizes an appropriate loss function over all the transliteration corpora.

$$M^* \quad = \quad \arg\min_{M} \mathcal{L}(M, \mathcal{D}) \tag{8.1}$$

where, M is the candidate joint transliteration model and $\mathcal{D}$=$(D_1, D_2, ..., D_l)$ is training data for all language pairs, $\mathcal{L}$ is the training loss function given the model and the training data.

We address three training scenarios of practical importance:

Similar Source Languages We have multiple orthographically similar *source* languages. There is a single target language which is not similar to the source languages. This is an instance of many-to-one learning, *e.g.*, Indic languages to English.

Similar Target Languages We have multiple orthographically similar *target* languages. There is a single source language which is not similar to the target languages. This is an instance of one-to-many learning, *e.g.*, English to Indic languages.

All Similar Languages We have multiple *source as well as target* languages, which are all orthographically similar. This is an instance of many-to-many learning, *e.g.*, Indic-Indic languages.

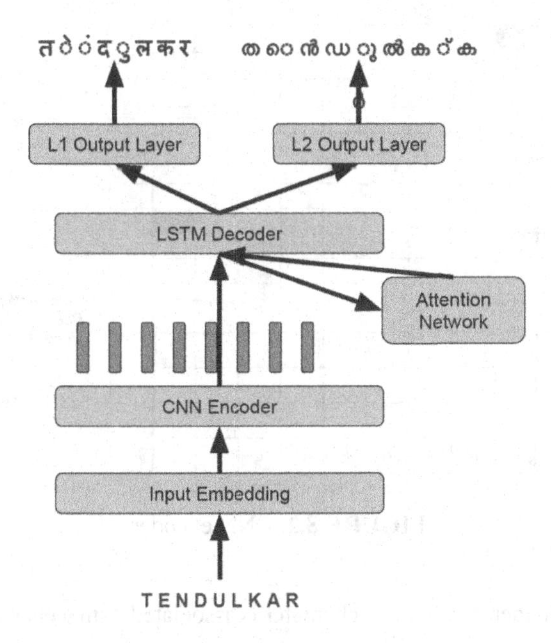

FIGURE 8.1: Multilingual neural architecture.

8.3.2 Network architecture

In this section, we describe our solution for multilingual transliteration. Our goal is to transliterate a word $\mathbf{x}^s$ in language s to its corresponding word $\mathbf{y}^t$ in the language t. This can be modelled as finding the word $\mathbf{y}^{t*}$ that maximizes the probability $p(\mathbf{y}^t|\mathbf{x}^s)$. Given that $\mathbf{x}^s = x_1^s, x_2^s, ...x_n^s$ and $\mathbf{y}^t = y_1^t, y_2^t, ...y_m^t$ are sequences of characters, we can further decompose these as:

$$\mathbf{y}^{t*} = \arg\max_{\mathbf{y}^t} \prod_{j=1}^{m} p(y_j^t|y_{j-1}^t...y_1^t, \mathbf{x}^s) \tag{8.2}$$

We model $\mathbf{P^{s,t}} = p(y_j^t|y_{j-1}^t...y_1^t, \mathbf{x}^s)$ using a encoder-decoder neural network. Note that we design a single network to represent all the $P^{s,t}$ distributions corresponding to the set of language pairs D. Our goal has been to develop a compact architecture where most of the components can be shared across languages. Figure 8.1 show the architecture of the network and the remainder of this section describes the details:

8.3.2.1 Input representation and character embedding

We represent each character in the input sequence by a onehot vector (x^s). The input vocabulary is the union of the character sets of all the source languages. Since the source languages are orthographically similar, there is a high degree of overlap

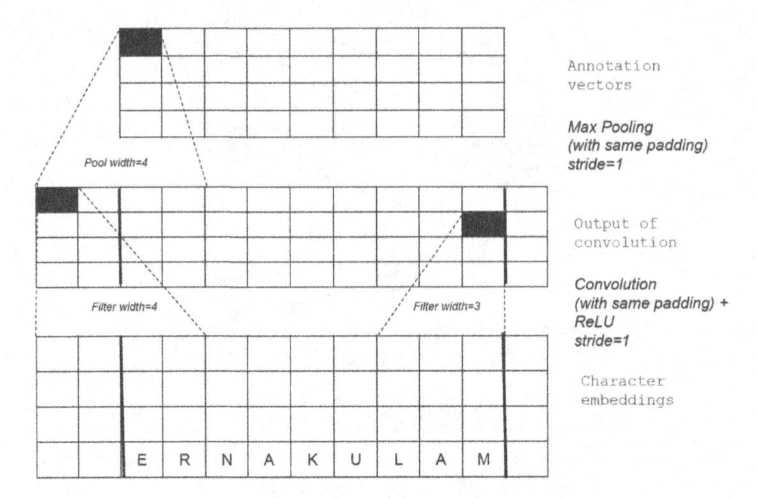

FIGURE 8.2: CNN encoder.

between their character sets. Each character is associated with a character embedding ($\Phi^S(x^s)$).

8.3.2.2 Encoder

Each character sequence generates a sequence of character embeddings. We used a CNN encoder, which is shared across all source languages, to encode this sequence. The CNN encoder consists of a single convolutional layer, with a stride size of 1 and SAME padding. This is followed by a ReLU non-linearity operation and max pooling. We use filters of different sizes (1 to 4) and concatenate their outputs to generate the output annotation vectors (a_i), one for each element in the input sequence. Figure 8.2 depicts the encoder layer.

$$a_1, ..., a_n = \text{Enc}_{CNN}([\Phi^S(x_1^s), ..., \Phi^S(x_n^s)]) \tag{8.3}$$

We chose the CNN encoder over the conventional bidirectional LSTM encoder since the temporal dependencies for transliteration are mostly local, which can be handled by the CNN encoder. We observed that this speeds up training and decoding significantly, without much impact on accuracy. Our CNN encoder architecture is simple compared to the CNN encoders used in machine translation, where the need to model long-term temporal dependencies required more involved architectures (Gehring et al., 2017; Lee et al., 2017).

8.3.2.3 Decoder

Following encoding of the input sequence, the decoder generates the output sequence, with one output vector generated in each time step. We use a layer of LSTM cells for decoding, whose start state is the average of the encoder's output vectors

(Sennrich et al., 2017). The decoder predicts the next output (o_j) based on the previous decoder state (h_{j-1}), the previous character output by the decoder (y^t_{j-1}) and a context vector (c_j).

$$h_0 = \frac{1}{n}\sum_{i=1}^{n} a_i \tag{8.4}$$

$$o_j, h_j = \text{LSTM}(h_{j-1}, [c_j, \Phi^T(y^t_{j-1})]) \tag{8.5}$$

The context vector (c_j) represents information from the input sequence relevant to the prediction of the next character and is generated by an attention mechanism.

8.3.2.4 Attention mechanism

The attention mechanism serves as a method for the decoder to take into consideration all encoder annotation vectors of the input sequence during computation for the next state. It does this by generating a context vector (c_j) which is a linear interpolation of the annotation vectors. The interpolation weights (α_{ij}) are determined as a function of the annotation vector and the decoder's previous state (h_{j-1}) and output (y_{j-1}). We use a small feedforward network (weights W^a and bias b^a) to learn this function. The attention mechanism is shared by all language pairs. The network input is the vector $v_{ij} = [a_i, h_{j-1}, \Phi^T(y^t_{j-1})]$.

$$e_{ij} = \tanh(W^a v_{ij} + b^a) \tag{8.6}$$

$$\alpha_{ij} = \frac{exp(e_{ij})}{\sum_{i=0}^{i=n} exp(e_{ij})} \tag{8.7}$$

$$c_j = \sum_{i=0}^{i=n} \alpha_{ij} a_i \tag{8.8}$$

8.3.2.5 Output layer

The LSTM's output (o_j) is transformed to the size of the output vocabulary, and a softmax function is applied to convert the output scores to probabilities. Each target language has its set of output layer parameters (weights $\mathbf{W^t}$ and bias $\mathbf{b^t}$ for language *t*).

$$s_j = \text{softmax}(\mathbf{W^t} o_j + \mathbf{b^t}) \tag{8.9}$$

$$y^{t*}_j = \arg\max_k s^k_j \tag{8.10}$$

Except for the output layer, all the other network parameters are shared across all languages. This allows maximum transfer of information for multilingual learning, while the output layer alone specializes for the specific target language. This avoids duplication of decoder parameters for each target language in previously proposed architectures. Compared to using a target language tag in the input sequence (Johnson et al., 2017), we think our approach allows the language parameters to directly influence the output characters.

8.3.2.6 Training objective

For each language pair (s,t), we can define the negative log-likelihood loss function ($L^{(s,t)}$). The objective of training the multilingual transliteration model is to minimize the negative log-likelihood loss for all language pairs:

$$L(\theta) = \frac{1}{l} \sum_{(s,t) \in \mathcal{D}} L^{(s,t)}(\theta) \tag{8.11}$$

where, θ refers to the network parameters.

8.3.2.7 Model selection

For each language pair, we selected the model with the maximum accuracy on the validation set. After training the model for a fixed number of iterations (sufficient for convergence), we select the model with the maximum accuracy on the validation set for each language pair. For instance, if the model corresponding to the 32nd epoch reported maximum accuracy on the validation set for English-Hindi, this model was used for reporting test set results for English-Hindi. We observed that this criterion performed better than choosing the model with least validation set loss over all language pairs.

8.4 Experimental Setup

We describe our experimental setup.

8.4.1 Network details

The CNN encoder has four filters (widths 1 to 4) of 128 hidden units each in the convolutional layer (encoder output size = 512). We use stride size of 1 and SAME padding for the convolutional and max-pooling layers. The decoder is a single layer of 512 LSTM units. We used the same configuration for both bilingual and multilingual experiments across all datasets for convenience after exploration on some language pairs, since hyperparameter tuning on a per dataset or language pair basis would be very time consuming. The maximum sequence length we considered was 30. We apply dropout (Srivastava et al., 2014) (probability = 0.5) at the outputs of the encoder and decoder, and SGD with ADAM optimizer (Kingma and Ba, 2014)

eng-Indic		**Indic-eng**		**Indic-Indic**				**ara-Slavic**	
eng-hin	12K	hin-eng	18K		ben	kan	tam	ara-ces	15K
eng-ben	13K	ben-eng	12K	hin	3620	5085	5290	ara-pol	15K
eng-kan	10K	kan-eng	15K	ben		2720	2901	ara-slv	10K
eng-tam	10K	tam-eng	15K	kan			4216	ara-slk	10K

TABLE 8.1: Training set statistics for different datasets (number of word pairs).

(learning rate = 0.001). We trained our models for a maximum of 40 epochs (which we found sufficient for our models to converge) and a batch size of 32. In each training epoch, we cycle through the parallel corpora of each language pair. The parallel corpora are roughly of the same size. Better training schedules like proportional mixing of examples from different languages pairs in each epoch could be explored in future.

8.4.2 Baseline PBSMT system

We also trained a bilingual transliteration system based on phrase-based SMT (PBMST) for comparison. The PBSMT system was trained using *Moses* (Koehn et al., 2007) with no lexicalized reordering and uses monotonic decoding. We used a 5-gram character language model trained with Witten-Bell smoothing.

8.4.3 Languages

We experimented with two sets of orthographically similar languages:

Indian languages (i) Hindi (*hin*) and Bengali (*ben*) from the Indo-Aryan branch of Indo-European family, (ii) Kannada (*kan*) and Tamil (*tam*) from the Dravidian family. We studied Indic-Indic transliteration and transliteration involving a non-Indian language (English↔Indic). We mapped equivalent characters in different Indic scripts in order to build a common vocabulary based on the common offsets of the Unicode codepoints. English uses the Latin script which is alphabetic.

Slavic languages Czech (*ces*), Polish (*pol*), Slovenian (*slv*) and Slovak (*slk*). We studied Arabic↔Slavic transliteration. Arabic is non-Slavic (Semitic branch of Afro-Asiatic) and uses an *abjad* script in which vowel diacritics are omitted in general usage.

The languages chosen are representative of languages spoken by some major groups of peoples: Indic, Romance, Germanic, Slavic, *etc.* These languages are spoken by around two billion people. So our approach addresses a major chunk of the world's people.

Pair	Src	Tgt
English-Indic		
eng-hin	KANAKLATA	कनकलता (*kanakalata*)
eng-kan	LEHMANN	ಲೆಹಮನ್ (*l.ehaman*)
Slavic-Arabic		
pol-ara	DUMITRESCU	دوميترسكو (*dwmytrskw*)
ces-ara	MAURICE	موريس (*mwrys*)

TABLE 8.2: Examples of transliteration pairs from our datasets.

8.4.4 Datasets

Indian languages We used the official *NEWS 2015* shared task dataset (Banchs et al., 2015) for English to Indic transliteration. This dataset has been used for many editions of the NEWS shared tasks. We split the *NEWS 2015* training dataset as the train and validation data for Indic-English transliteration. For testing, we used the *NEWS 2015* dev-test set. We created the Indian-Indian parallel transliteration corpora from the English to Indian language training corpora of the NEWS 2015 dataset by mining name pairs which have English names in common.

Slavic languages We mined the Arabic-Slavic dataset from *Wikidata* (Vrandečić and Krötzsch, 2014), a structured knowledge base containing *items* (roughly entities of interest). Each item has a *label* (title of item page) which is available in multiple languages. We extracted labels from selected items referring to named entities (persons, organizations and locations) to ensure that we extract parallel transliterations (as opposed to translations). We identified the relevant items by filtering on the basis of the coarse hierarchy into which the items are organized via *instance of* links.

Table 8.2 shows some examples from the two datasets. Table 8.1 shows training set statistics for the datasets used in our experiments. The validation set contains 1000 word pairs for English→Indic and Arabi-Slavic language pairs, 500 word pairs for Indic→English and Indic-Indic language pairs. The test set contains 1000 word pairs for all language pairs.

8.4.5 Evaluation

We use top-1 exact match accuracy as the evaluation metric (Banchs et al., 2015). This is one of the metrics in the NEWS shared tasks on transliteration.

8.5 Results and Discussion

We discuss/analyze the results of our experiments.

Pair	P	B	M	Pair	P	B	M
Similar Source and Target Languages							
Indic-Indic (45.5%)							
ben-hin	**29.74**	19.08	27.69	kan-ben	28.59	24.04	**37.47**
ben-kan	17.62	18.14	**27.74**	kan-tam	34.89	30.85	**38.30**
hin-ben	29.92	25.46	**39.15**	tam-hin	**29.07**	19.24	28.97
hin-tam	25.15	28.62	**38.70**	tam-kan	26.99	19.86	**29.06**
Similar Source Languages							
Slavic-Arabic (55.8%)				*Indic-English (24.2%)*			
ces-ara	38.91	37.10	**59.17**	ben-eng	**55.23**	48.93	54.01
pol-ara	34.70	34.80	**44.83**	hin-eng	49.19	38.26	**51.11**
slk-ara	43.26	37.49	**62.21**	kan-eng	42.79	33.77	**47.70**
slv-ara	41.90	36.74	**62.04**	tam-eng	**33.93**	23.22	25.93
Similar Target Languages							
Arabic-Slavic (176.8%)				*English-Indic (1.1%)*			
ara-ces	15.41	12.08	**36.76**	eng-ben	42.90	41.70	**46.10**
ara-pal	13.68	12.26	**24.21**	eng-hin	60.50	**64.10**	60.70
ara-slk	15.24	13.82	**38.72**	eng-kan	48.70	52.00	**53.90**
ara-slv	18.31	13.63	**44.35**	eng-tam	52.90	**57.80**	55.30

TABLE 8.3: Comparison of bilingual (B) and multilingual (M) neural models as well as bilingual PBSMT (P) models (top-1 accuracy %).

8.5.1 Quantitative observations

Table 8.3 compares results of bilingual (B) and multilingual (M) neural models as well as a baseline bilingual transliteration system based on phrase-based statistical machine transliteration (P).

We observe that multilingual training substantially improves accuracy over bilingual training in all datasets (an average increase of 58.2% over all language pairs). Transliteration accuracy increases in all scenarios: (i) similar sources, (ii) similar targets and (iii) similar sources and targets.

If we look at results for various language groups, transliteration involving Slavic languages and Arabic benefits more than transliteration involving Indic languages and English. Arabic→Slavic transliteration shows maximum improvement (average: 176.8%) while English→Indic pairs show minimum improvement (average: 1.1%).

We also see that the multilingual model shows significant improvements over a bilingual transliteration system based on phrase-based SMT.

The PBSMT system is better than the bilingual neural transliteration system in most cases. This is consistent with previous work (Finch et al., 2015), where the standard encoder-decoder architecture (with attention mechanism) (Bahdanau et al., 2015) could not outperform PBSMT approaches. However, Finch et al. (2016)'s model, which uses target bidirectional LSTM with model ensembling, outperforms PBSMT models. These improvements are orthogonal to our work and could be used to further improve the bilingual as well as multilingual systems. The small size of the datasets and the shallowness of the networks could be potential reasons why the neural network models are not able to outperform the PBSMT approach.

8.5.2 Qualitative observations

We see that multilingual transliteration is better than bilingual transliteration in the following aspects:

- Vowels are generally a major source of transliteration errors (Kumaran et al., 2010; Kunchukuttan and Bhattacharyya, 2015) because of ambiguities in vowel mappings. We see a major decrease in vowel errors due to multilingual training (average decrease of ∼20%). We observe a substantial decrease in long-short vowel confusion errors (Indic languages as target languages) and insertion/deletion of *A* (English/Slavic as target). We also see a major improvement in Arabic→Slavic transliteration. The Arabic script does not represent vowels, hence the transliteration system needs to correctly generate vowels. The multilingual model is better at generating vowels compared to the bilingual model.

- We also observe that consonants with similar phonetic properties are a major source of transliteration errors, and these show a substantial decrease in multilingual training. For Indic-English transliteration, we see substantial error reduction in the following character pairs *K-C*, *T-D*, *P-B*. We also observe a decrease in confusion between aspirated and unaspirated sounds. For Arabic→Slavic transliteration, we see substantial error reduction for the following character pairs *K-C*, *F-V* and *P-B*.

- For Slavic→Arabic, we observed a significant reduction in the number of errors related to characters representing fricative sounds like *j,s,z,g*.

- The multilingual system seems to prefer the canonical spellings, even when other alternative spellings seem faithful to the source language phonetics. It is thus able to learn conventional usage better than the bilingual models, *e.g.*, *morisa* (Hindi, romanized text is shown) is transliterated incorrectly to the phonetically acceptable English word *moris* by the bilingual model. The multilingual model generates the correct system *Maurice*.

- Since Indic scripts are very phonetic, very few non-canonical spellings are possible. As a consequence, vowel error reduction was also minimum for English-Indic transliteration (10%). This may partly explain why multilingual training provides minimal benefit for English-Indic transliteration.

Table 8.4 shows some examples where the multilingual output is better than the bilingual output.

Pair	Source	B	M
ara-ces	باستور (*bAstwr*)	bastor	pastor
ara-ces	كيلين (*kylyn*)	kelen	kailin
hin-eng	वर्जि (*varjila*)	vergill	virgil
hin-eng	एलिसन (*elisana*)	elissan	ellison

TABLE 8.4: Examples of bi- *vs.*, multi-lingual outputs. *ara* and *hin* text are also shown using Buckwalter and ITRANS romanization schemes, respectively.

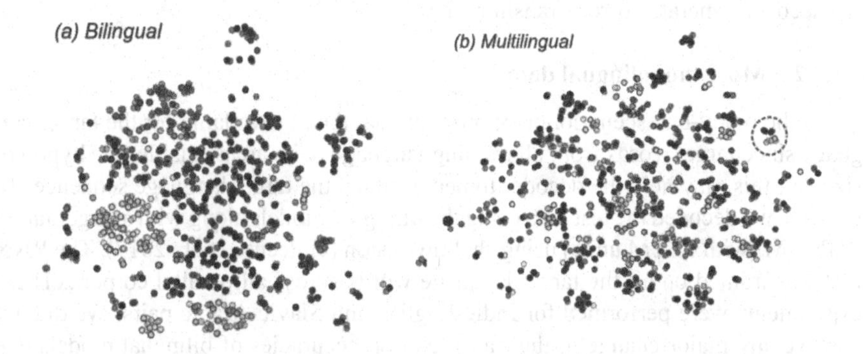

FIGURE 8.3: Visualization of contextual representations of vowels for *hin-eng* transliteration. Each colour represents a different vowel.

8.5.3 Analysis

We investigated few hypotheses to understand why multilingual models are better:

8.5.3.1 Better contextual representations for vowels

We hypothesize that the encoder learns better contextual representations for vowels. To test this hypothesis, we studied three character long sequences from the test set with a vowel in the middle (*i.e.,* 1-char window around vowel). We processed these sequences through the encoder of the bilingual and multilingual transliteration systems to generate the encoder output corresponding to the vowels. *e.g.,* For the vowel *a* in the word *part*, we encode the three character sequence *par* using the encoder. The encoder output corresponding to the character *a* is considered the contextual representation of the character *a* in this word. Figure 8.3 shows a visualization of these contextual representations of the vowels using t-SNE (van der Maaten and Hinton, 2008). For the bilingual model, we observe that the contextual representations of same vowels tend to cluster together. For the multilingual model, the clustering is more specialized. The representations are grouped by the vowel along with the context. For instance, the region highlighted in the plot shows representations of Hindi vowels *e* (white) and *i* (black) followed by the consonant *v*. Other vowels with the same con-

Pair	B	M
pol-ara	14.64	44.83
hin-eng	38.26	35.93

TABLE 8.5: Results of experiments under balanced data conditions.

text are seen in the same region too. This suggests that the multilingual model is able to learn specialized representations of vowels in different contexts and this may help the decoder generate correct transliterations.

8.5.3.2 More monolingual data

In the many-one scenario, more monolingual data is available for the target language since target words from all training language pairs are available. We hypothesize that this may help the decoder to better model the target language sequence. To test this, we decoded the test data using the bilingual models along with a larger target RNN LM (with LSTM units) using shallow fusion (Gulcehre et al., 2017). The RNN LM was trained on all the target language words across all parallel corpora. These experiments were performed for Indic-English and Slavic-Arabic pairs. We did not observe any major change in the transliteration accuracies of bilingual models due to the integration of a larger LM. Thus, larger target side data does not explain the improvement in transliteration accuracy due to multilingual transliteration.

8.5.3.3 More parallel data

Multilingual training pools together parallel corpora from multiple orthographically similar languages, which effectively increases the data available for training. To test if this explains the improved performance, we compared multilingual and bilingual models under similar data size conditions *i.e.,* the total parallel corpora size across all language pairs used to train the multilingual model is equivalent to the size of the bilingual corpus used to train the bilingual model. Specifically, we compared the following under similar data size conditions: (a) *{ben,hin,kan,tam} — eng* multilingual model (50% *hin-eng* training pairs) with *hin-eng* bilingual model, (b) *{ces,pol,slk,slv} — ara* multilingual model (30% *pol-ara* training pairs) with *pol-ara* bilingual model. Table 8.5 shows the results of these experiments. We observed that the multilingual system showed significantly higher transliteration accuracy compared to the bilingual model for Polish-English. For Hindi-English, the bilingual model was better than the multilingual model. So, we cannot decisively conclude that the performance improvement of multilingual transliteration can be attributed to the effective increase in training data size. In a multilingual training scenario, data from other languages act as a noisy version of data from the original language and supplement the available bilingual data. But they cannot necessarily substitute data from the original language pair.

Pair	$\mathbf{sep}_{dec}$	$\mathbf{sep}_{out}$	$\mathbf{sep}_{none}$	Pair	$\mathbf{sep}_{dec}$	$\mathbf{sep}_{out}$	$\mathbf{sep}_{none}$
Indic-Indic							
ben-hin	27.28	27.69	**28.72**	kan-ben	32.22	**37.47**	35.76
ben-kan	25.86	**27.74**	27.11	kan-tam	40.06	38.30	**40.37**
hin-ben	37.22	39.15	**39.35**	tam-hin	27.74	**28.97**	28.45
hin-tam	35.74	**38.70**	35.54	tam-kan	28.13	**29.06**	28.65
Arabic-Slavic				*English-Indic*			
ara-ces	**39.68**	36.76	35.95	eng-ben	41.00	**46.10**	44.10
ara-pol	**27.25**	24.21	26.85	eng-hin	56.00	60.70	**61.60**
ara-slk	**41.36**	38.72	40.14	eng-kan	49.30	53.90	**54.30**
ara-slv	**49.14**	44.35	45.88	eng-tam	53.00	**55.30**	52.90

TABLE 8.6: Comparison of multilingual architectures.

8.5.4 Comparison with variant architectures

We compare three variants of the encoder-decoder architecture:

- **Separate output layer** (sep_{out}): This is our proposed model, where there is an output layer (and parameters) for every target language. The decoder is shared across all languages.
- **Separate decoder** (sep_{dec}): Every target language has its own decoder and output layer (Lee et al., 2017).
- **No separate layers** (sep_{none}): All languages share the same decoder and output layer, but the first token of the input sequence is a special token to specify target language. This is the architecture used by Google's NMT system (Johnson et al., 2017).

These architectures differ in the degree of parameter sharing. sep_{dec} has fewer shared parameters than our model and sep_{none} has more shared parameters than our model. In all three cases, the encoder is shared across all source languages. Table 8.6 shows the results of this comparison. We cannot definitively conclude if one model is better than the other. Except for Arabic-Slavic transliteration, the trend seems to indicate that models with greater parameter sharing (sep_{dec}/sep_{none}) may perform better. In any case, given the comparable results, we prefer models with fewer model parameters (sep_{dec}/sep_{none}).

8.6 Zeroshot Transliteration

In the previous sections, we have shown that multilingual training is beneficial for language pairs observed during training. In addition, the encoder-decoder architecture opens up the possibility of *zeroshot transliteration i.e.,* transliteration between language pairs that have not been seen during training. The encoder-decoder architecture decouples the source and target language network components and makes the architecture more modular. As a consequence, we can consider the encoder output (for the source language) to be embedded in a language neutral, common subspace — a sort of interlingua. The decoder proceeds to generate the target word from the language-neutral representation of the source word. Hence, training on just a few language pairs is sufficient to learn all language-specific parameters — making zeroshot transliteration possible.

Before describing different zeroshot transliteration scenarios, we introduce a few terms. A language that is the source in any language pair seen during training is said to be *source-covered.* We can define *target-covered* languages analogously. Now, we can envisage the following zeroshot transliteration scenarios:

- *Unseen language pairs*: both the source and target languages are covered, but the pair was not observed during training
- *Unseen source language*: the source language is not covered, but it is orthographically similar to other source-covered languages
- *Unseen target language*: can be defined analogously.

Next, we describe our proposed solutions to these scenarios.

8.6.1 Unseen language pair

We investigated the following solutions:

8.6.1.1 Multilingual zeroshot-direct

Since the source and target languages are covered, we use the trained multilingual model discussed in previous sections for source-target transliteration.

Model selection can be an issue for this approach. As discussed earlier, model selection using validation set accuracy for each language pair is better than average validation set loss. For an unseen language pair, we cannot compute the validation set accuracy/loss for model selection (since validation data is not available). So we explored the following model selection criterion: maximum average validation set accuracy across all the trained language pairs (*sc_acc*). We also compared *sc_acc* to the model with least average validation set loss over all training language pairs (*sc_loss*).

Nevertheless, there are inherent limitations to model selection by averaging validation accuracy or loss for trained language pairs. Irrespective of the model selection method used, the chosen model may still be suboptimal for unseen language pairs since the network is not optimized for such pairs.

8.6.1.2 Multilingual zeroshot-pivot

To address the limitations with direct transliteration, we propose transliteration from source to target using a pivot language, and pipelining the best source-pivot and pivot-target transliteration models. We choose a pivot language such that the network has been trained for source-pivot and pivot-target pairs. Since the network has been trained for the source-pivot and target-pivot pairs, we can expect optimal performance in each stage of the pipeline. *Note that we use the multilingual model for source-pivot and pivot-target transliteration* (we found it better than using the bilingual models). To reduce cascading errors due to pipelining, we consider the top-k source-pivot transliterations in the next stage of the pipeline. The target word $\mathbf{t}$ for a source word $\mathbf{s}$ is computed as:

$$P(\mathbf{t}|\mathbf{s}) \approx \sum_{i=1}^{k} P(\mathbf{t}|\mathbf{p}^i)P(\mathbf{p}^i|\mathbf{s}) \tag{8.12}$$

$\mathbf{p}^i$: i^{th} best source-pivot transliteration. We used $k = 10$.

8.6.2 Unseen source language

An unseen source language can be easily handled in our architecture. Though the language has not been source-covered, a source word can be directly processed through the network since all source languages share the encoder and character embeddings.

8.6.3 Unseen target language

Handling an unseen target language is tricky, since the output layer is specific to each target language. Hence, parameters for the unseen language's output layer cannot be learnt during training. Note that even architectures where the entire network is completely shared between all language pairs (Johnson et al., 2017) cannot handle an unseen target language — the embedding for target language indicator tokens are not learnt during training for unseen target languages. We found that simple approaches like assigning parameters for unseen target languages by averaging the parameters of the trained target languages do not work.

Hence, we use a target-covered language as a proxy for the unseen target language. The simplest approach considers the output of the source to proxy language transliteration system as the target language's output. However, this doesn't take into account the phonotactic characteristics of the target language. We propose to incorporate this information by using an RNN (using LSTM units) character-level language model of the target language words. While predicting the next output character during decoding, we combine the scores from the multilingual transliteration model and the target LM using shallow fusion of the transliteration model and the language model (Gulcehre et al., 2017).

Pair	Biling Direct	Zeroshot Pivoting	Zeroshot Direct		
			sc_acc	*sc_loss*	*sc_ora*
ben-tam	16.20	**36.12** (hin)	31.79	27.45	34.47
tam-ben	18.48	**47.01** (hin)	24.87	17.97	25.28
hin-kan	34.66	**47.14** (tam)	44.48	42.13	43.05
kan-hin	34.12	**39.02** (tam)	40.45	37.28	40.96

TABLE 8.7: Results for zeroshot transliteration: Unseen language pair. Best pivot for multilingual pivoting in brackets.

8.6.4 Results and discussion

8.6.4.1 Unseen language pair

We experimented with transliteration between four Indic languages *viz. hin,ben, tam,kan.* We trained a multilingual model on eight out of the 12 possible language pairs, covering all the four languages. The remaining four language pairs (*tam-ben, ben-tam, hin-kan* and *hin-kan*) are the unseen language pairs (results in Table 8.7).

Zeroshot *vs.*, Direct Bilingual: For all unseen language pairs, all zeroshot systems (pivot as well as different direct configurations) are better than the direct bilingual system. Note that unlike the zeroshot systems, the bilingual systems were directly trained on the unseen language pairs. Yet the zeroshot systems outperform direct bilingual systems since the underlying multilingual models are significantly better than the bilingual models (as seen in Section 8.5). These results show that the multilingual model generalizes well to unseen language pairs.

Direct *vs.*, Pivot Zeroshot: We also observe that the pivot zeroshot system is better than both the direct zeroshot systems (*sc_acc* and *sc_loss*). To verify if the limitations of the model selection criterion explain the direct system's relatively lesser accuracy, we also considered an oracle direct zeroshot system (*sc_ora*). The oracle system selects the model with the best accuracy using a parallel validation corpus for the unseen language pair. This oracle system is also inferior to the pivot transliteration model. So, we can conclude that the network is better tuned for transliteration of language pairs it has been directly trained on. Hence, multilingual pivoting works better than direct transliteration in spite of the cascading errors involved in pipelining.

Model Selection Criteria: For the direct zeroshot system, the average accuracy (*sc_acc*) is a better model selection criterion than average loss (*sc_loss*).

8.6.4.2 Unseen source language

We conducted 2 experiments: (a) train on *(ben,kan,tam)-eng* pairs and test on *hin-eng* pair, (b) train on *(pol,slk,slv)-ara* pairs and test on *ces-ara* pair. See Table 8.8 for

Method	Slavic-ara *(ces-ara)*	Indic-eng *(hin-eng)*
Bilingual	37.10	38.26
Zeroshot	**60.48**	45.24
Multilingual	59.17	**51.11**

TABLE 8.8: Results for zeroshot transliteration: Unseen source language.

Method	ara-Slavic *(ara-ces)*		eng-Indic *(eng-hin)*	
	Proxy	acc	Proxy	acc
Bilingual	none	12.08	none	**64.10**
	slk	42.09	kan	9.00
Proxy	slv	41.89	ben	18.30
	pol	38.27	tam	2.90
Proxy	slk	**42.20**	kan	9.70
+	slv	40.99	ben	18.10
LMFusion	pol	36.35	tam	3.10

TABLE 8.9: Results for zeroshot transliteration: Unseen target language.

results. In this scenario too, we observe that zeroshot transliteration performs better than direct bilingual transliteration. In fact, zeroshot transliteration is competitive with multilingual transliteration too (accuracy is $>$ 90% of multilingual transliteration). Though the network has not seen the source language, the encoder is able to generate source language representations that are useful for decoding.

8.6.4.3 Unseen target language

We conducted two experiments: (a) train on *ara-(pol,slk,slv)* pairs and test on *ara-ces* pair, (b) train on *eng-(ben,kan,tam)* pairs and test on *eng-hin* pair. The target languages used in training were the proxy languages. See Table 8.9 for results.

We observe contradictory results from the experiments. For *ara-ces*, the use of proxy language gives transliteration results better than bilingual transliteration. But, the proxy language is not a good substitute for Hindi. Shallow fusion of target LM with the transliteration model makes little difference.

We see that the transliteration performance of proxy languages is correlated to its orthographic similarity to the target language. Thus, it is preferable to choose a proxy with high orthographic similarity to the target language. We see one anomaly to this trend. Kannada as proxy performs badly compared to Bengali, though Kannada is orthographically more similar to Hindi. One reason could be the orthographic convention in Hindi and Bengali that the terminal vowel is automatically suppressed. In Kannada, the vowel has to be explicitly suppressed with a terminal *halanta* character. Simply deleting the terminal *halanta* in Kannada output to conform to Hindi conventions increases accuracy to 24.1% (better than Bengali). Clearly shallow fusion is not sufficient to adapt a proxy language's output to the target language, and further investigations are required. If a proxy-target corpus is available, we can generate better transliterations via pivoting.

8.7 Incorporating Phonetic Information

So far, we considered characters as atomic units. We have thus relied on correspondences between characters for multilingual learning. For some languages, however, we can find an almost one-one correspondence from the characters to phonemes (a basic unit of sound). Each phoneme can be factorized into a set of articulatory properties like the place of articulation, nasalization, voicing, aspiration *etc.* If the input for transliteration incorporates these phonetic properties, it may learn better character representations across languages by bringing together similar characters. *e.g.*, the Kannada character ಳ (*La*), has no Hindi equivalent character, but the Hindi character ल (*la*) is the closest character. The two characters differ in terms of one phonetic feature (the *retroflex* property), which can be represented in the phonetic input and can serve to indicate the similarity between the two characters.

We incorporated phonetic features in our model by using feature-rich input vectors instead of the conventional onehot vector input for characters. Our phonetic feature input vector is a bitvector encoding the phonetic properties of the character, one bit for each value of every property. The multiplication of the phonetic feature vector with the weight matrix in the first layer generates phonetic embeddings for each character. These are inputs to the encoder. Apart from this input change, the rest of the network architecture is the same as described in Section 8.3.2.

8.7.1 Experiments

We experimented with Indian languages (Indic→English and Indic-Indic transliteration). Indic scripts generally have a one-one correspondence from characters to phonemes. Hence, we use he phonetic features described by Kunchukuttan et al. (2016) to phonetic feature vectors for characters. These Indic languages are spoken by nearly a billion people and hence the use of phonetic features is useful for many of the world's most widely spoken languages.

8.7.2 Results and discussion

Table 8.10 shows the results. We observe that phonetic feature input improves transliteration accuracy for Indic-English transliteration. The improvements are primarily due to reduction in errors related to similar consonants like (T,D), (P,B), (C,K) and the use of H for aspiration.

	Pair	O	P^h	Pair	O	P^h
Indic-English	ben-eng	54.01	**55.53**	kan-eng	47.70	**53.31**
	hin-eng	51.11	**54.86**	tam-eng	25.93	**29.63**
Indic-Indic	ben-hin	27.69	**28.51**	kan-ben	37.47	**39.19**
	ben-kan	27.74	**29.72**	kan-tam	38.30	**41.93**
	hin-ben	**39.15**	37.73	tam-hin	28.97	**29.17**
	hin-tam	38.70	**39.00**	tam-kan	29.06	**31.54**

TABLE 8.10: Onehot (O) *vs.*, phonetic (P^h) input.

For Indic-Indic transliteration, we see moderate improvement in transliteration accuracy due to phonetic feature input. Since the source, as well as target scripts, are largely phonetic, phonetic representation may not be useful in resolving ambiguities (unlike Indic-English transliteration). Again, we see improvements due to reduction of errors related to similar consonants.

8.8 Summary and Future Work

We propose a compact neural encoder-decoder model for multilingual transliteration involving orthographically similar languages, that is designed to ensure maximum sharing of parameters across languages while providing room for learning language-specific parameters. This allows greater sharing of knowledge across language pairs by leveraging orthographic similarity.

The following are the contributions of our work:

1. We show that multilingual transliteration exhibits significant improvement in transliteration accuracy over bilingual transliteration in different scenarios (average improvement of 58%). Our results are backed by extensive experiments on eight languages across two orthographically similar language groups.

2. We perform an error analysis which suggests that representations learnt by the encoder in multilingual transliteration can reduce transliteration ambiguities. Multilingual transliteration also seems better at learning canonical transliterations instead of alternative, phonetically equivalent transliterations. These could explain the improved performance of multilingual transliteration.

3. We empirically show that models with maximal parameter sharing are beneficial, without increasing the model size.

4. We explore the *zeroshot transliteration task* (*i.e.,* transliteration between languages/language pairs not seen during training) and show that our multilingual model can generalize well to unseen languages/language pairs. Notably, the

zeroshot transliteration results mostly outperform the direct bilingual transliteration model. Multilingual pivot transliteration performs better than multilingual direct translation.

5. We have richer phonetic information at our disposal for some related languages. We propose a novel method to incorporate phonetic input in the model and show that it provides modest gains for multilingual transliteration.

Transliteration is an example of a sequence to sequence task which is characterized by the following properties:

- small vocabulary
- short sequences
- monotonic transformation
- unequal source and target sequence length
- significant vocabulary overlap across languages.

Given the benefits we have shown for multilingual transliteration, other NLP tasks that can be characterized similarly (*viz.* grapheme to phoneme conversion, translation of short text like tweets and headlines between related languages at subword level) could also benefit from multilingual training.

Chapter 9

Conclusion and Future Directions

In this chapter, we summarize the work done in this monograph, describe the conclusions from the work and discuss possible future work.

9.1 Conclusions

In this monograph, we addressed two sequence transformation tasks between related languages: machine translation and transliteration. These tasks are key to enabling cross-lingual communication and breaking down linguistic barriers. They are particularly important for related languages since people speaking related languages generally reside in contiguous geographical areas and communicate heavily amongst themselves for administrative, business and social needs.

Many of these related languages have few linguistic resources or parallel corpora. The broad context of the work is the exploration of solutions in scenarios where resources required for developing translation/transliteration systems (*e.g.,* parallel translation corpora, parallel transliteration corpora, morph-analyzers, *etc.*) are limited. We seek methods which improve translation quality and transliteration accuracy in such low-resource scenarios, enable easy sharing and portability of resources by utilizing language relatedness.

This is important to ensure that digital information is available in many languages at an affordable price. Translation and transliteration services between related languages can serve the information needs of a large user population across the globe. High quality solutions need huge investments for every language; something that may not be possible for many individuals, corporations, institutions and governments. It is therefore infeasible to serve large number of languages using this approach. Most of the translation and transliteration services are between English and other languages, and translation/transliteration between related languages is supported via pivoting only in most cases.

We considered languages that can trace their relatedness to the following sources:

- Genetic relatedness (creating language families like Indo-European and subfamilies like Indo-Aryan)
- Contact relatedness (creating linguistic areas like the Indian subcontinent and the Balkans)

We explored methods to utilize language relatedness for translation and transliteration at different linguistic levels. We utilized the following similarities of related languages to improve translation and transliteration:

- Lexical similarity
- Morphological isomorphism
- Structural correspondence
- Orthographic similarity

From our work, **we conclude that it is beneficial to take language relatedness into account for building machine translation and transliteration systems in limited resource scenarios**. The benefits accrue by:

- Using lexical similarity and multilingual learning to alleviate the knowledge acquisition bottleneck (in this context, it means inability to learn translation mappings for most of the vocabulary due to limited training data).
- Using linguistic similarities to share linguistic resources.

While it is nearly impossible to learn machine translation/translation (or for that matter any other NLP task) in a completely language independent method, related languages provide a useful level of abstraction — a level at which partial language independence (between the related languages) can be achieved to a great extent.

The following subsections summarize the work presented and major findings reported in the monograph for machine translation and transliteration between related languages.

9.1.1 Machine translation

Utilizing lexical similarity between related languages through the use of subword-level representations is a major theme in our work on translation between related languages.

9.1.1.1 Linguistically and statistically informed subword representation for SMT is beneficial

We find that character and character n-grams do not have the representational power to model translation between related languages (unless they are very close). Hence, we propose the use of two variable length subwords units as basic translation units between related languages in order to utilize lexical similarity between these languages: orthographic syllables and byte pair encoded units. While orthographic syllables are linguistically motivated pseudo-syllables, BPE identifies the most frequent character sequences as basic units based on statistical properties of the text. We find that the improvement in translation quality due to OS and BPE units can be attributed to the following reasons:

- Reduction in data sparsity compared to word and morpheme translation units.
- Judicious use of lexical similarity, balancing the use of word-level information with use of subword level information.
- Ability to learn diverse lexical mappings, with the ability to translate cognates as well as non-cognates.

We show that these subword units are beneficial in resource-scarce scenarios also:

- Parallel corpus is very limited.
- Cross-domain translation.

9.1.1.2 Enriching pivot and multilingual translation with subword representation is beneficial

We find that when multilingual learning is complemented by subword-level representation, the resulting translation quality in a resource-constrained scenario can match a resource-rich scenario. For instance, subword-level pivot-based SMT models are competitive with the best direct translation systems. In addition, we find that a pivot-based SMT model with multiple related pivot languages and subword-level representation is nearly equivalent to a direct SMT model. Thus, the use of subwords as translation units coupled with multiple related pivot languages can compensate for the lack of a direct parallel corpus.

9.1.1.3 Reusing resources among related languages is possible and beneficial

We find significant benefit in reusing resources created for one language for other related languages. We show this by a case study of simply reusing the pre-ordering rules for English-Hindi PBSMT for other Indian languages. This is a simple approach, but points to the potential for reuse among related languages, and the possibility of more sophisticated approaches which can provide further gains.

9.1.1.4 Findings from case study for translation involving Indian languages

We conduct an extensive case-study on translation involving nine major Indian languages covering 72 language pairs. We experiment with the different translation units discussed previously. The following are the major findings:

1. The case study provides further evidence that OS and BPE-level translation models perform significantly better than word and morpheme level models.
2. Based on translation quality, we see a clear partitioning of translation pairs by language family. For instance, translations involving Indo-Aryan languages can be done with a high level of accuracy, whereas those involving Dravidian languages are extremely difficult. This suggests that SMT approaches customized to language family pairs may be investigated.

3. Rich morphology of Indian languages, especially Dravidian languages, is a major factor impacting translation quality. For instance, it is easiest to translate to/from Hindi (a language with a relatively isolating morphology). On the other hand, translation involving Malayalam (a highly agglutinative language) is the most difficult.

9.1.2 Machine transliteration

We focus on challenging transliteration tasks where language relatedness is crucial: *unsupervised transliteration* and *multilingual transliteration*. We find that utilization of orthographic similarity between languages leads to substantial improvement in the transliteration accuracy for both tasks. Orthographic similarity enables establishing initial character-level transliteration correspondences and ensures that all related languages have similar character language modelling behaviour.

9.1.2.1 Substring-based unsupervised transliteration

The following are our findings:

- Without appropriate inductive bias, it is not possible to learn a reasonably good transliteration system in an unsupervised setting. Hence, we proposed a character-level model using an Expectation Maximization with MAP estimation (EM-MAP) framework. The inductive bias is provided through prior distributions on the model parameters. These prior distributions capture phonetic similarity between characters. The phonetic knowledge is drawn from the orthographic similarity between the languages.

- Character-level learning is not sufficient, contextual information is also needed for disambiguation of transliterations. We propose a substring-based model bootstrapped from the character-based model to incorporate contextual information.

- The top-10 accuracy of our unsupervised system is comparable with the top-1 accuracy of supervised systems; hence it can be used in downstream applications like MT and cross-lingual information retrieval which can consume top-k lists or confusion networks generated by the transliteration system and perform further disambiguation.

9.1.2.2 Multilingual transliteration

The following are our findings:

- Training multiple transliteration systems involving related languages can be modelled as a multitask learning problem, where every task corresponds to a language pair. This is a consequence of the orthographic similarity between the languages.

- To ensure maximum utilization of task relatedness, we propose a neural encoder-decoder architecture which maximizes parameter sharing across languages. We find that these compact networks, with far fewer model parameters, perform as well as networks with specialized components and more model parameters.
- We find that the multilingual transliteration models significantly outperform bilingual transliteration models.
- We find that representations learnt by the encoder in the multilingual transliteration model can reduce transliteration ambiguities. Multilingual transliteration also seems better at learning canonical transliterations instead of alternative, phonetically equivalent transliterations. These could explain the improved performance of multilingual transliteration compared to bilingual transliterations.
- We find the network can generalize to languages/language pairs not encountered during training. This is exemplified by the competitive performance of zeroshot transliteration (*i.e.,* transliteration between languages/language pairs not seen during training) compared to supervised, bilingual transliteration.

9.2 Future Work and Directions

This monograph explores a few questions/problems and proposes some solutions regarding translation and transliteration involving related languages. This section discusses future directions emanating from this work. Our hope is that future advances in technology can help develop cost-effective and high-quality solutions for translation and transliteration involving related languages. While current solutions target translation between English and related languages, we hope to see enough advances to support translation and transliteration among related languages too. In addition, dialects of a language could also be supported. We group these future directions of work into four categories: Statistical Machine Translation, Neural Machine Translation, Machine Transliteration and NLP for related languages.

9.2.1 Statistical machine translation

Our work dealt with various scenarios related to subword-level translation and multilingual learning; there are a few more directions that can be explored.

9.2.1.1 Subword-level translation

- Using Byte Pair Encoding, the subword vocabulary is learnt to maximize a monolingual objective (data likelihood/compression). Ideally, we would like to learn the vocabulary (and consequently the segmentation of words) to maximize a bilingual objective (translation quality or a surrogate objective). Previous work on bilingual segmentation with reference to languages which have no delimiters may be a good starting point (Xu et al., 2008; Huang et al., 2008; Ma and Way, 2009; Paul et al., 2011), but these methods may need to be adapted or new methods developed to take into account lexical similarity.

- Translation models trained on different translation units may have their own strengths and weaknesses. Hence, one way to improve the translation system could be to combine these different translation models. For instance, we can combine a word-level model and BPE-level model. Since these models are trained at various granularities, the easiest way is to combine the outputs of each MT system *i.e.,* generate a new MT output by interleaving partial outputs from different systems. Previous work has attempted to combine word and character-level models with modest success (Nakov and Tiedemann, 2012). We performed a few initial studies combining various configurations of word, morpheme, OS and BPE-level models, but these did not yield any major improvements. We believe that further exploration with alternative methods like reranking candidates from different systems, combining at the phrase-table level, *etc.* may be promising directions to explore.

- We have experimented extensively with Indian languages. These cover many of the major languages of the world spoken by more than a billion speakers (including 8 out of the 20 most widely spoken languages in the world). Hence, our results on the effectiveness of subword-level translation are widely applicable. We have also experimented with representative language pairs from other language families like Slavic, Polynesian, Altaic, *etc.* However, further experimentation with a larger set of these related language groups will only serve to further validate our results and potentially throw up interesting observations.

- The work in this monograph was done in the broader context of resource-scarce related languages. Nevertheless, it would be interesting to study how subword-level translation compares with word-level translation when large parallel corpora are available.

- One of the challenges to the practical utility of subword translation is the increase in alignment and decoding time due to increase in the length of sentence on account of subword-level representation. While we have made progress in improving decoding time, further improvements can be explored. Faster methods for subword-level alignment also need to be explored in order to improve training time and speed-up experimental iterations.

- Given the success of subword-level models for related languages, it would be logical to see if subword-level models or other methods to utilize lexical similarity can be useful in improving the performance of machine translation-related tasks like domain-adaptation, automatic post-editing, translation quality estimation, translation evaluation, *etc.*

- In this monograph, we have only addressed supervised translation between related languages. When parallel corpus is not available between two languages, we have explored pivot translation. In the event that these avenues are not feasible, unsupervised translation approaches using only monolingual corpora can be explored. Since the languages are related, we can make appropriate assumptions and model similarities between the languages involved to provide inductive bias to drive the learning process. A starting point for this direction is work on unsupervised translation (not necessarily restricted to related languages) (Ravi and Knight, 2011; Dou and Knight, 2012, 2013; Dou et al., 2014). Some recent work on unsupervised translation between related languages may also be relevant (Pourdamghani and Knight, 2005). We could also explore unsupervised translation involving multiple related languages. The fact that the multiple languages are related may provide a stronger inductive bias which can partially compensate for the lack of parallel corpora.

9.2.1.2 Multilingual learning

- For pivot translation, we have only considered a related pivot language. However, a relevant scenario is that the source and target languages may share parallel corpora with an unrelated pivot language. Even though the source and target languages are related, the methods we proposed cannot leverage lexical similarity while pivoting via an unrelated language. Hence, better methods to leverage source-target relatedness would be an interesting question to explore.
- An important and interesting translation use case that needs further research is: pivot language related to source or target language, but not both. This scenario arises in translation from a set of related languages to an unrelated *lingua franca*. Simple solutions like using a pipeline approach with subword-level translation between related languages and word-level translation between the unrelated pair provides only limited benefit. Word order divergence is a major problem for such solutions.

9.2.2 Neural machine translation

In recent years, there is an increasing volume of work which shows that neural machine translation performs better than traditional statistical machine translation. The encoder-decoder architecture is the most popular NMT architecture. Automatic feature learning, distributed representation, continuous space modelling, end-to-end training are some prominent advantages of NMT. However, current NMT systems require large-scale parallel corpora for good performance (Koehn and Knowles, 2017). In contrast, our work on machine translation between related languages is useful for low-resource settings. However, advances in transfer learning have shown that NMT can work well in low-resource scenarios also by transferring knowledge from high resource languages.

Recently massively multilingual NMT models have been built which are trained on parallel corpora of all major language pairs of the world. Such networks have a large number of parameters, and can suffer from language interference. This approach contrasts with language pair specific models. In this case, gains from multilinguality may be limited and models have to be deployed for every language pair. Most of the gains for multilingul translation translation come from related language. Hence, a middle path might be to build multilingual models that cater to a set of related languages. This can result in smaller networks while improving translation quality.

Most literature has focussed only on major languages (100 odd languages). Transfer learning using related languages can help extend MT systems to support minority languages. Some work has been done in this direction, but research is in a nascent stage in this area.

Pivot, zeroshot and unsupervised NMT have helped translation between languages which do not have a parallel corpus. It would be an interesting research direction to see if these methods can be further improved for related languages by utilizing their language relatedness.

Many of the potential future directions we discussed in the context of SMT also applies to NMT *e.g.,* bilingual segmentation, unsupervised translation, domain adaptation, *etc.*

9.2.3 Machine transliteration

As discussed previously, we view the transliteration task as a simplified version of the task of translation between related languages. Hence, we would like to explore and adapt methods and learning from transliteration to translation between related languages.

There are a few possible directions to extend our work on transliteration between related languages:

Unsupervised Transliteration In our work, we have considered unsupervised transliteration between two languages. We could extend this to unsupervised learning of transliteration models for multiple related languages. The fact that the multiple languages are related may provide a stronger inductive bias which can partially compensate for the lack of parallel transliteration corpora.

Multilingual Transliteration Since multilingual transliteration has shown significant benefits, it would be worth investigating multilingual training for other NLP tasks that can be characterized similarly *e.g.,* grapheme to phoneme conversion, translation of short text like tweets and headlines between related languages at subword level.

9.2.4 An NLP pipeline for related languages

We have seen that related languages form a useful layer of abstraction for building translation and transliteration systems. An interesting direction of exploration is the development of an NLP pipeline for related languages. Such a pipeline should be able to share corpora and resources from all related languages. It should also be able to learn systems that are more generalizable by learning from multiple related languages.

The ideal scenario would be a single model that works across all related languages for a particular task. For instance, the pipeline could contain models shared across different related languages for tasks in the NLP pipeline like POS Tagging, chunking, parsing, semantic role labelling, named entity recognition, *etc.* On top of that, we could train specific applications like document classification, sentiment analysis, question/answering, *etc.* that are shared across languages too. Such multilingual models would be able to support languages for which the system has not been explicitly trained.

Traditionally, translation has been seen as the most demanding consumer of the language processing pipeline, which thrives on advances in other NLP tasks. However, in the scenario we have envisioned above, translation is a foundational technology to enable building multilingual systems. Translation (or an appropriate process to learn bilingual mappings) and transliteration are required to establish representations across multiple languages before joint learning and sharing of resources can fructify.

Appendices

Appendix A

Extended ITRANS Romanization Scheme

The extended ITRANS notation provides a scheme for transcription of major Indian scripts in Roman, using ASCII characters. It extends the ITRANS transliteration scheme to cover characters not covered in the original scheme. Tables A.1a shows the ITRANS mappings for vowels and diacritics (*matras*). Table A.1b shows the ITRANS mappings for consonants. These tables also show the Unicode offset for each character. By Unicode offset, we mean the offset of the character in the Unicode range assigned to the script. For Indic scripts, logically equivalent characters are assigned the same offset in their respective Unicode codepoint ranges. For illustration, we also show the Devanagari characters corresponding to the transliteration.

NOTES:

1. The ITRANS scheme has the same character for the independent vowel and the corresponding vowel diacritic (*matra*).
2. For consonants, the Roman character represents the consonant without the implicit *schwa* vowel. For instance, *k* represents क्. To represent the implicit vowel, *a* has to follow the consonant (*ka* represents क).

ITRANS	Unicode Offset	Devanagari
a	05	अ
aa, A	06, 3E	आ, ा
i	07, 3F	इ, ि
ii, I	08, 40	ई, ी
u	09, 41	उ, ु
uu, U	0A, 42	ऊ, ू
RRi, R^i	0B, 43	ऋ, ृ
RRI, R^I	60, 44	ॠ, ॄ
LLi, L^i	0C, 62	ऌ, ॢ
LLI, L^I	61, 63	ॡ, ॣ
.e	0E, 46	ऎ, ॆ
e	0F, 47	ए, े
ai	10, 48	ऐ, ै
.o	12, 4A	ऒ, ॊ
o	13, 4B	ओ, ो
au	14, 4C	औ, ौ
aM	05 02, 02	अं
aH	05 03, 03	अः
.m	02	ं
.h	03	ः

(a) Vowels

ka 15 क	kha 16 ख	ga 17 ग	gha 18 घ	~Na, N^a 19 ङ
cha 1A च	Cha 1B छ	ja 1C ज	jha 1D झ	~na, JNa 1E ञ
Ta 1F ट	Tha 20 ठ	Da 21 ड	Dha 22 ढ	Na 23 ण
ta 24 त	tha 25 थ	da 26 द	dha 27 ध	na 28 न
pa 2A प	pha 2B फ	ba 2C ब	bha 2D भ	ma 2E म
ya 2F य	ra 30 र	la 32 ल	va, wa 35 व	
sha 36 श	Sha 37 ष	sa 38 स	ha 39 ह	
Ra 31 ऱ	lda, La 33 ळ	zha 34 ऴ		

(b) Consonants

Appendix B

Software and Data Resources

In the course of this research, we have created software and resources for NLP spanning related languages (particularly for machine translation). Most of these resources are targeted towards Indian languages. The software is made available through the open-source GPL 3.0[1] License and the data resources are made available under the Creative Commons (Attribution-NonCommercial-ShareAlike)[2] license.

B.1 Software

1. *Indic NLP Library*: This is a Python library containing NLP components for multiple Indian languages covering normalizer, tokenizer, transliterator, script converter, word segmenter, querying script information, computing phonetic similarity, orthographic syllabification, *etc.* The goal of the library is to build tools that work across multiple Indian languages by leveraging the similarities among these languages.

   URL: `http://anoopkunchukuttan.github.io/indic_nlp_library`

   URL: `https://github.com/anoopkunchukuttan/meteor_indic`

2. *CFILT Pre-order*: This is a fork of the source reordering rules for English-Hindi SMT developed by Ramanathan et al. (2008) and Patel et al. (2013). We have refactored the code and provided a user-friendly command-line interface. While the rules were originally written for English-Hindi translation, we have verified that these rules are useful for translation from English to all major Indian languages.

   URL: `https://github.com/anoopkunchukuttan/cfilt_preorder`

3. *IIT Bombay Unsupervised Transliterator*: Unsupervised transliteration system which uses phonetic features to define transliteration priors. This is an EM-based method which builds on the work of (Ravi and Knight, 2009) and is an implementation of our work described in Chapter 7.

   URL: `https://github.com/anoopkunchukuttan/transliterator`

[1] `https://www.gnu.org/licenses/gpl-3.0.en.html`
[2] `https://creativecommons.org/licenses/by-nc-sa/4.0`

4. *Multilingual Neural Machine Transliteration System*: A multilingual Neural Machine Transliteration system which can jointly train multiple language pairs (possibly related). It is an implementation of our work described in Chapter 8.

   URL: `https://github.com/anoopkunchukuttan/mlxlit`

B.2 Online Systems

1. *Shata-Anuvaadak*: This online system provides statistical machine translation services via a user interface for 110 Indian language pairs. The system currently hosts word-level phrase-based SMT systems. We plan to host subword-level SMT systems for Indian languages soon, as well as providing REST APIs to access the translation service.

   URL: `http://www.cfilt.iitb.ac.in/indic-translator`

2. *BrahmiNet*: This online system provides the following services for 306 language pairs of the Indian subcontinent: (a) statistical transliteration, (b) script conversion between Indic scripts, and (c) romanization of Indic scripts (using the extended ITRANS transliteration scheme we proposed). We provide a user interface as well as a REST API to access this service.

   URL: `http://www.cfilt.iitb.ac.in/brahminet`

B.3 Data Resources

1. *IIT Bombay English-Hindi Parallel Corpus*: The corpus is a compilation of parallel corpora previously available in the public domain as well as new parallel corpora we collected. The corpus contains 1.49 million parallel segments, of which 694k segments were not previously available in the public domain. This corpus has been used in two editions of shared tasks at the Workshop on Asian Language Translation (2016 and 2017). To the best of our knowledge, this is the largest publicly available English-Hindi parallel corpus. Other English-Indian language corpora are much smaller than this corpus. Hence, this corpus could be used for improving English to Indian language machine translation, where Hindi acts as a resource-rich language for other Indian languages.

   URL: `http://www.cfilt.iitb.ac.in/iitb_parallel`

2. *Translation Resources for 110 Indian Language Pairs*: It consists of trained phrase-based SMT models for 110 Indian language pairs. These can be used for

(a) research into post-editing methods, (b) building pivot-based SMT methods, (c) build translation applications like a mobile app, multi-lingual chat, *etc.*

URL: `http://www.cfilt.iitb.ac.in/~moses/shata_anuvaadak/register.html`

3. *Parallel Transliteration Corpora for 110 Indian Language Pairs*: These corpora were mined from parallel translation corpora using the Moses Transliteration Module (Durrani et al., 2014). It contains around 1.7 million word pairs across all language pairs — roughly 15,000 word pairs per language pair. It can be used for building transliteration systems as well as studying cognates, loanwords and similar phenomena in Indian languages.

   URL: `http://www.cfilt.iitb.ac.in/brahminet/static/register.html`

Appendix C

Conferences/Workshops for Translation between Related Languages

The following is a listing of specific conferences and workshops that cater to NLP for related languages. Obviously, the major conferences and journals in Natural Language Processing and Computational Linguistics address this topic but these specialized forums complement them and this nascent research area. The following is a partial listing of the major venues for research on NLP for related languages:

- Workshop on NLP for Similar Languages, Varieties and Dialects (VarDial)
- Workshop on Language Technology for Closely Related Languages and Language Variants (LT4CloseLang)
- Workshop on South and Southeast Asian Natural Languages Processing (WS-SANLP)
- International Conference on Natural Language Processing (ICON): Although the conference agenda is not restricted to related languages, ICON tends to have a significant representation of work in Indian language NLP since it is the premier NLP conference in India.
- International Joint Conference on Natural Language Processing (IJCNLP): Although the conference agenda is not restricted to related languages, IJCNLP is a good source on literature related to Asian languages since it is the premier NLP conference in Asia.
- Nordic Conference on Computational Linguistics (NoDaLiDa): The aim of the conference is to bring together researchers from the five Nordic countries to discuss all aspects of language technology.
- Workshop on Balto-Slavic NLP (BSNLP)
- ACM Transactions on Asian and Low-Resource Language Information Processing (TALLIP): It is a journal dedicated to NLP in Asian languages, low-resource languages of Africa, Australasia, Oceania and the Americas.

In addition, some tutorials have addressed MT for related languages:

1. Anoop Kunchukuttan, Mitesh Khapra, Pushpak Bhattacharyya. *Statistical Machine Translation between Related Languages*. North American Chapter of the Association for Computational Linguistics. 2016.

2. Anoop Kunchukuttan, Mitesh Khapra. *Translation and Transliteration between Related Languages*. International Conference on Natural Language Processing. 2015.

Bibliography

Abbi, A. (2012). Languages of India and India and as a Linguistic Area. `http://www.andamanese.net/LanguagesofIndiaandIndiaasalinguisticarea.pdf`. Retrieved November 15, 2015.

Agrawal, R. (2017). Towards efficient Neural Machine Translation for Indian Languages. Master's thesis, International Institute of Information Technology Hyderabad.

Akmajian, A., Farmer, A. K., Bickmore, L., Demers, R. A., and Harnish, R. M. (2017). *Linguistics: An introduction to language and communication*. MIT Press.

Al-Onaizan, Y., Curin, J., Jahr, M., Knight, K., Lafferty, J., Melamed, D., Och, F., Purdy, D., Smith, N., and Yarowsky, D. (1999). Statistical Machine Translation. Technical report, Johns Hopkins University.

Al-Onaizan, Y. and Knight, K. (2002). Translating named entities using monolingual and bilingual resources. In *Proceedings of the 40th Annual Meeting on Association for Computational Linguistics*, pages 400–408.

Anthes, G. (2010). Automated Translation of Indian Languages. *Communications of the ACM*, 53(1):24–26.

Atreya, A., Chaudhari, S., Bhattacharyya, P., and Ramakrishnan, G. (2016). Value the Vowels: Optimal Transliteration Unit Selection for Machine. In *Unpublished, private communication with authors*.

Bahdanau, D., Cho, K., and Bengio, Y. (2015). Neural machine translation by jointly learning to align and translate. In *International Conference on Learning Representations*.

Banchs, R. E., Zhang, M., Duan, X., Li, H., and Kumaran, A. (2015). Report of NEWS 2015 machine transliteration shared task. In *Proceedings of the Fifth Named Entities Workshop*, page 10.

Banerjee, S. and Lavie, A. (2005). METEOR: An automatic metric for MT evaluation with improved correlation with human judgments. In *Proceedings of the ACL Workshop on Intrinsic and Extrinsic Evaluation Measures for Machine Translation and/or Summarization*, volume 29, pages 65–72.

Begum, R., Husain, S., Dhwaj, A., Sharma, D. M., Bai, L., and Sangal, R. (2008). Dependency Annotation Scheme for Indian Languages. In *International Joint Conference on Natural Language Processing*, pages 721–726.

Bender, E. M. (2011). On achieving and evaluating language-independence in NLP. *Linguistic Issues in Language Technology*, 6(3):1–26.

Bharati, A., Chaitanya, V., Kulkarni, A. P., Sangal, R., and Rao, G. U. (2003). Anusaaraka: Overcoming the Language Barrier in India. *Anuvad: Approaches to Translation.*

Bharati, A., Chaitanya, V., and Sangal, R. (1996). *Natural language processing: A Paninian perspective.* Prentice-Hall of India.

Bharati, A., Gupta, M., Yadav, V., Gali, K., and Sharma, D. M. (2009). Simple parser for Indian languages in a dependency framework. In *Proceedings of the Third Linguistic Annotation Workshop*, pages 162–165. Association for Computational Linguistics.

Bhat, R. A. (2017). *Exploiting Linguistic Knowledge to Address Representation and Sparsity Issues in Dependency Parsing of Indian Languages.* PhD thesis, International Institute of Information Technology Hyderabad.

Bisani, M. and Ney, H. (2008). Joint-sequence models for grapheme-to-phoneme conversion. *Speech communication*, 50(5), 434-451.

Bisazza, A., Ruiz, N., Federico, M., and Kessler, F.-F. B. (2011). Fill-up versus interpolation methods for phrase-based SMT adaptation. In *International Workshop on Spoken Language Translation*, pages 136–143.

Bojar, O., Diatka, V., Rychlý, P., Straňák, P., Suchomel, V., Tamchyna, A., and Zeman, D. (2014). HindEnCorp — Hindi-English and Hindi-only Corpus for Machine Translation. In *Proceedings of the 9th International Conference on Language Resources and Evaluation.*

Brown, P., Pietra, V. J. D., Pietra, S. A. D., and Mercer, R. L. (1993). The mathematics of statistical machine translation: Parameter estimation. *Computational linguistics*, 19(2), 263-311.

Bynon, T. (1977). *Historical linguistics.* Cambridge University Press.

Caruana, R. (1997). Multitask learning. *Machine learning*, 28(1), 41-75.

Chang, M.-W., Goldwasser, D., Roth, D., and Tu, Y. (2009). Unsupervised constraint driven learning for transliteration discovery. In *Proceedings of Human Language Technologies: The 2009 Annual Conference of the North American Chapter of the Association for Computational Linguistics.*

Cherry, C. and Foster, G. (2012). Batch tuning strategies for statistical machine translation. In *Proceedings of the 2012 Conference of the North American Chapter of the Association for Computational Linguistics: Human Language Technologies.*

Chiang, D. (2007). Hierarchical phrase-based translation. *Computational Linguistics*, 33(2), 201-228.

Chinnakotla, M. K., Damani, O. P., and Satoskar, A. (2010). Transliteration for resource-scarce languages. *ACM Transactions on Asian Language Information Processing (TALIP)*, 9(4), 1-30.

Chitnis, R. and DeNero, J. (2015). Variable-length word encodings for neural translation models. In *Proceedings of the 2015 Conference on Empirical Methods in Natural Language Processing*.

Chung, J., Cho, K., and Bengio, Y. (2016). A character-level decoder without explicit segmentation for neural machine translation. In *Proceedings of the Meeting of the Association for Computational Linguistics*.

CICEKLI, K. A. E. I. (2002). A machine translation system between a pair of closely related languages. In *Proceedings of the 17th International Symposium on Computer and Information Sciences*, pages 192–196.

Cohn, T. and Lapata, M. (2007). Machine translation by triangulation: Making effective use of multi-parallel corpora. In *Proceedings of the Meeting of the Association for Computational Linguistics*.

Collins, M., Koehn, P., and Kučerová, I. (2005). Clause restructuring for statistical machine translation. In *Annual meeting on Association for Computational Linguistics*.

Collobert, R., Weston, J., Bottou, L., Karlen, M., Kavukcuoglu, K., and Kuksa, P. (2011). Natural language processing (almost) from scratch. *Journal of Machine Learning Research*, 12(Aug):2493–2537.

Covington, M. (1996). An algorithm to align words for historical comparison. *Computational linguistics*, 22(4), 481-496.

Dabre, R., Cromiers, F., Kurohashi, S., and Bhattacharyya, P. (2015). Leveraging small multilingual corpora for SMT using many pivot languages. In *Proceedings of the 2015 Conference of the North American Chapter of the Association for Computational Linguistics: Human Language Technologies*.

Dabre, R., Kunchukuttan, A., Fujita, A., and Sumita, E. (2018). NICT's participation in WAT 2018: Approaches using multilingualism and recurrently stacked layers. In *Proceedings of the 5th Workshop on Asian Language Translation*, Hong Kong, China.

Dabre, R., Nakagawa, T., and Kazawa, H. (2017). An Empirical Study of Language Relatedness for Transfer Learning in Neural Machine Translation. In *The 31st Pacific Asia Conference on Language, Information and Computation*.

Ding, C., Utiyama, M., and Sumita, E. (2016). Similar Southeast Asian Languages: Corpus-Based Case Study on Thai-Laotian and Malay-Indonesian. In *Workshop on Asian Language Translation*, page 149.

Dong, D., Wu, H., He, W., Yu, D., and Wang, H. (2015). Multi-Task Learning for Multiple Language Translation. In *Annual Meeting of the Association for Computational Linguistics*.

Dou, Q. and Knight, K. (2012). Large scale decipherment for out-of-domain machine translation. In *Proceedings of the 2012 Joint Conference on Empirical Methods in Natural Language Processing and Computational Natural Language Learning*, pages 266–275.

Dou, Q. and Knight, K. (2013). Dependency-based decipherment for resource-limited machine translation. In *Proceedings of the Conference on Empirical Methods in Natural Language Processing*, pages 1668–1676.

Dou, Q., Vaswani, A., and Knight, K. (2014). Beyond parallel data: Joint word alignment and decipherment improves machine translation. In *Proceedings of the Conference on Empirical Methods in Natural Language Processing*, pages 557–565.

Durrani, N., Hoang, H., Koehn, P., and Sajjad, H. (2014). Integrating an unsupervised transliteration model into Statistical Machine Translation. In *Proceedings of the Conference of the European Chapter of the Association for Computational Linguistics*.

Durrani, N. and Hussain, S. (2010). Urdu word segmentation. In *Proceedings of Human Language Technologies: The 2010 Annual Conference of the North American Chapter of the Association for Computational Linguistics*.

Durrani, N., Sajjad, H., Fraser, A., and Schmid, H. (2010). Hindi-to-Urdu machine translation through transliteration. In *Proceedings of the 48th Annual Meeting of the Association for Computational Linguistics*.

Ekbal, A., Naskar, S. K., and Bandyopadhyay, S. (2006). A modified joint source-channel model for transliteration. In *Proceedings of the COLING/ACL on Main Conference Poster Sessions*.

Emeneau, M. B. (1956). India as a linguistic area. *Language*, 32(1), 3–16.

Finch, A., Liu, L., Wang, X., and Sumita, E. (2015). Neural network transduction models in transliteration generation. In *Proceedings of the Fifth Named Entities Workshop*.

Finch, A., Liu, L., Wang, X., and Sumita, E. (2016). Target-bidirectional neural models for machine transliteration. *Proceedings of The Sixth Named Entities Workshop (NEWS)*.

Finch, A. M. and Sumita, E. (2010). A Bayesian model of bilingual segmentation for transliteration. In *International Workshop on Spoken Language Translation*.

Firat, O., Cho, K., and Bengio, Y. (2016a). Multi-way, multilingual neural machine translation with a shared attention mechanism. In *Conference of the North American Chapter of the Association for Computational Linguistics*.

Firat, O., Sankaran, B., Al-Onaizan, Y., Vural, F. T. Y., and Cho, K. (2016b). Zero-resource translation with multi-lingual neural machine translation. In *Conference on Empirical Methods in Natural Language Processing*.

Gage, P. (1994). A New Algorithm for Data Compression. In *The C Users Journal*, 12(2), 23–38.

Galušcáková, P. and Bojar, O. (2012). Improving SMT by using parallel data of a closely related language. In *Proceedings of HLT: The Baltic Perspective*, pages 58–65.

Ganapathiraju, M., Balakrishnan, M., Balakrishnan, N., and Reddy, R. (2005). Om: One tool for many (Indian) languages. *Zhejiang University Science*, 6(11):1348.

Gehring, J., Auli, M., Grangier, D., and Dauphin, Y. N. (2017). A Convolutional Encoder Model for Neural Machine Translation. In *International Conference on Learning Representations*.

Gillick, D., Brunk, C., Vinyals, O., and Subramanya, A. (2016). Multilingual Language Processing from Bytes. In *Conference of the North American Chapter of the Association for Computational Linguistics*.

Gispert, A. D. and Marino, J. B. (2006). Catalan-english statistical machine translation without parallel corpus: Bridging through Spanish. In *Proceedings of 5th International Conference on Language Resources and Evaluation (LREC)*.

Goldhahn, D., Eckart, T., and Quasthoff, U. (2012). Building Large Monolingual Dictionaries at the Leipzig Corpora Collection: From 100 to 200 Languages. In *Language Resources and Evaluation Conference*, pages 759–765.

Goldwater, S. and McClosky, D. (2005). Improving statistical MT through morphological analysis. In *Proceedings of the Conference on Human Language Technology and Empirical Methods in Natural Language Processing*, pages 676–683.

Gulcehre, C., Firat, O., Xu, K., Cho, K., and Bengio, Y. (2017). On integrating a language model into neural machine translation. *Computer Speech & Language*, 45, 137–148.

Hajič, J., Hric, J., and Kuboň, V. (2000). Machine translation of very close languages. In *Proceedings of the Sixth Conference on Applied Natural Language Processing*.

Haspelmath, M. (2001). The European linguistic area: Standard Average European. In Haspelmath, M., editor, *Language typology and language universals: An international handbook*. Walter de Gruyter.

He, J., Wu, L., Zhao, X., and Yan, Y. (2017). HCCL at SemEval-2017 Task 2: Combining multilingual word embeddings and transliteration model for semantic similarity. In *Proceedings of the 11th International Workshop on Semantic Evaluation (SemEval-2017)*, pages 220–225.

Heafield, K., Kayser, M., and Manning, C. D. (2014). Faster phrase-based decoding by refining feature state. In *Annual Meeting-Association For Computational Linguistics*, pages 130–135.

Heaps, H. S. (1978). *Information retrieval: computational and theoretical aspects*. Academic Press, Inc.

Hermjakob, U., Knight, K., and Daumé III, H. (2008). Name Translation in Statistical Machine Translation-Learning When to Transliterate. In *Conference of the Association for Computational Linguistics*, pages 389–397.

Hoang, H., Bogoychev, N., Schwartz, L., and Junczys-Dowmunt, M. (2016). Fast, Scalable Phrase-based SMT Decoding. *arXiv Pre-print arXiv:1610.04265*.

Homola, P. (2008). A hybrid machine translation system for typologically related languages. In *Proceedings of the 21st International Florida-Artificial-IntelligenceResearch-Society Conference, FLAIRS*.

Homola, P. and Kubon, V. (2004). A translation model for languages of acceding countries. In *Proceedings of the EAMT Workshop*.

Huang, C.-C., Chen, W.-T., and Chang, J. S. (2008). Bilingual Segmentation for Alignment and Translation. In *Computational Linguistics and Intelligent Text Processing: 9th International Conference*, pages 445–453.

Huang, L. and Chiang, D. (2007). Forest rescoring: Faster decoding with integrated language models. In *Annual Meeting-Association For Computational Linguistics*.

Hutchins, W. J. and Somers, H. L. (1992). *An introduction to machine translation*, volume 362. Academic Press London.

Inkpen, D., Frunza, O., and Kondrak, G. (2005). Automatic identification of cognates and false friends in French and English. In *Proceedings of the International Conference Recent Advances in Natural Language Processing*.

Irvine, A., Callison-Burch, C., and Klementiev, A. (2010). Transliterating from all languages. In *Proceedings of the Conference of the Association for Machine Translation in the Americas (AMTA)*.

Jagarlamudi, J. and Daumé III, H. (2012). Regularized interlingual projections: Evaluation on multilingual transliteration. In *Proceedings of the 2012 Joint Conference on Empirical Methods in Natural Language Processing and Computational Natural Language Learning*.

Jawaid, B., Kamran, A., and Bojar, O. (2014). Urdu Monolingual Corpus. LINDAT/CLARIN digital library at the Institute of Formal and Applied Linguistics, Charles University in Prague.

Jelinek, F. (1969). Fast sequential decoding algorithm using a stack. *IBM Journal of Research and Development*, 13(6):675–685.

Jha, G. N. (2012). The TDIL program and the Indian Language Corpora Initiative. In *Language Resources and Evaluation Conference.*

Jiampojamarn, S., Bhargava, A., Dou, Q., Dwyer, K., and Kondrak, G. (2009). DirecTL: A language-independent approach to transliteration. In *Proceedings of the 2009 Named Entities Workshop: Shared Task on Transliteration.*

Jiampojamarn, S., Cherry, C., and Kondrak, G. (2008). Joint Processing and Discriminative Training for Letter-to-Phoneme Conversion. In *Annual Meeting-Association for Computational Linguistics.*

Johnson, M., Schuster, M., Le, Q. V., Krikun, M., Wu, Y., Chen, Z., Thorat, N., Viégas, F., Wattenberg, M., Corrado, G., et al. (2017). Google's Multilingual Neural Machine Translation System: Enabling Zero-Shot Translation. *Transactions of the Association for Computational Linguistics.*

Karimi, S., Scholer, F., and Turpin, A. (2011). Machine transliteration survey. *ACM Computing Surveys (CSUR).*

Kashani, M. M., Joanis, E., Kuhn, R., Foster, G., and Popowich, F. (2007). Integration of an Arabic transliteration module into a statistical machine translation system. In *Proceedings of the Second Workshop on Statistical Machine Translation*, pages 17–24.

Khapra, M. M., Kumaran, A., and Bhattacharyya, P. (2010). Everybody loves a rich cousin: An empirical study of transliteration through bridge languages. In *Human Language Technologies: The 2010 Annual Conference of the North American Chapter of the Association for Computational Linguistics.*

Kingma, D. and Ba, J. (2014). Adam: A Method for Stochastic Optimization. In *International Conference on Learning Representations.*

Klementiev, A. and Roth, D. (2006). Weakly supervised named entity transliteration and discovery from multilingual comparable corpora. In *Proceedings of the 21st International Conference on Computational Linguistics and the 44th Annual Meeting of the Association for Computational Linguistics.*

Knight, K. (1999). Decoding complexity in word-replacement translation models. *Computational Linguistics*, 25(4):607–615.

Knight, K. and Graehl, J. (1998). Machine transliteration. *Computational Linguistics*, 24(4), 599–612.

Knight, K., Nair, A., Rathod, N., and Yamada, K. (2006). Unsupervised analysis for decipherment problems. In *Proceedings of the COLING/ACL on Main Conference Poster Sessions*. Association for Computational Linguistics.

Koehn, P. (2004). Statistical significance tests for machine translation evaluation. In *Proceedings of the Conference on Empirical Methods in Natural Language Processing.*

Koehn, P. (2005). Europarl: A parallel corpus for statistical machine translation. In *MT Summit*, volume 5, pages 79–86.

Koehn, P. (2009). *Statistical machine translation.* Cambridge University Press.

Koehn, P., Birch, A., and Steinberger, R. (2009). 462 machine translation systems for Europe. *Proceedings of MT Summit XII*, pages 65–72.

Koehn, P., Hoang, H., Birch, A., Callison-Burch, C., Federico, M., Bertoldi, N., Cowan, B., Shen, W., Moran, C., Zens, R., et al. (2007). Moses: Open source toolkit for statistical machine translation. In *Proceedings of the 45th Annual Meeting of the ACL on Interactive Poster and Demonstration Sessions.*

Koehn, P. and Knowles, R. (2017). Six challenges for neural machine translation. In *First Workshop on Neural Machine Translation.*

Koehn, P., Och, F. J., and Marcu, D. (2003). Statistical phrase-based translation. In *Proceedings of the 2003 Conference of the North American Chapter of the Association for Computational Linguistics on Human Language Technology-Volume 1*, pages 48–54. Association for Computational Linguistics.

Kondrak, G. (2000). A new algorithm for the alignment of phonetic sequences. In *Proceedings of the 1st North American chapter of the Association for Computational Linguistics Conference.*

Kondrak, G., Marcu, D., and Knight, K. (2003). Cognates can improve statistical translation models. In *Proceedings of the Conference of the North American Chapter of the Association for Computational Linguistics on Human Language Technology.*

Kumaran, A., Khapra, M. M., and Bhattacharyya, P. (2010). Compositional machine transliteration. *ACM Transactions on Asian Language Information Processing*, 9(4), 1–29.

Kunchukuttan, A. and Bhattacharyya, P. (2015). Data representation methods and use of mined corpora for Indian language transliteration. In *Named Entities Workshop.*

Kunchukuttan, A., Bhattacharyya, P., and Khapra, M. (2016). Substring-based Unsupervised Transliteration with Phonetic and Contextual Knowledge. In *SIGNLL Conference on Computational Natural Language Learning (CoNLL).*

Kunchukuttan, A., Puduppully, R., and Bhattacharyya, P. (2015). Brahmi-Net: A Transliteration and Script Conversion System for Languages of the Indian Subcontinent. In *Conference of the North American Chapter of the Association for Computational Linguistics — Human Language Technologies: System Demonstrations*.

Lee, J., Cho, K., and Hofmann, T. (2017). Fully character-level neural machine translation without explicit segmentation. *Transactions of the Association for Computational Linguistics*, 5, 365–378.

Lee, Y.-S. (2004). Morphological analysis for statistical machine translation. In *Proceedings of HLT-NAACL 2004: Short Papers*, pages 57–60.

Li, H., Kumaran, A., Pervouchine, V., and Zhang, M. (2009). Report of NEWS 2009 machine transliteration shared task. In *Proceedings of the 2009 Named Entities Workshop: Shared Task on Transliteration*.

Ma, Y. and Way, A. (2009). Bilingually motivated domain-adapted word segmentation for statistical machine translation. In *Proceedings of the 12th Conference of the European Chapter of the Association for Computational Linguistics*, pages 549–557.

Maimaiti, M., Liu, Y., Luan, H., and Sun, M. (2019). Multi-round transfer learning for low-resource NMT using multiple high-resource languages. *ACM Transactions on Asian Resource Language Information Processing*, 18(4):1–26.

Marchand, Y., Adsett, C. R., and Damper, R. I. (2009). Automatic syllabification in English: A comparison of different algorithms. *Language and Speech*, 52(1):1–27.

Melamed, I. D. (1995). Automatic evaluation and uniform filter cascades for inducing n-best translation lexicons. In *Third Workshop on Very Large Corpora*.

Moore, R. (2005). A discriminative framework for bilingual word alignment. In *Proceedings of the Conference on Human Language Technology and Empirical Methods in Natural Language Processing*.

More, R., Kunchukuttan, A., Dabre, R., and Bhattacharyya, P. (2015). Augmenting Pivot Based SMT with Word Segmentation. In *International Conference on Natural Language Processing*.

Nakov, P. and Ng, H. T. (2009). Improved statistical machine translation for resource-poor languages using related resource-rich languages. In *Proceedings of the 2009 Conference on Empirical Methods in Natural Language Processing*.

Nakov, P. and Tiedemann, J. (2012). Combining word-level and character-level models for machine translation between closely-related languages. In *Proceedings of the 50th Annual Meeting of the Association for Computational Linguistics*.

Neubig, G. and Hu, J. (2018). Rapid adaptation of neural machine translation to new languages. In *Proceedings of the 2018 Conference on Empirical Methods in Natural Language Processing*, pages 875–880, Brussels, Belgium. Association for Computational Linguistics.

Nguyen, T. Q. and Chiang, D. (2017). Transfer Learning across Low-Resource, Related Languages for Neural Machine Translation. In *International Joint Conference on Natural Language Processing.*

Och, F. J. (2003). Minimum error rate training in statistical machine translation. In *Proceedings of the 41st Annual Meeting on Association for Computational Linguistics-Volume 1*, pages 160–167. Association for Computational Linguistics.

Och, F. J. and Ney, H. (2002). Discriminative training and maximum entropy models for statistical machine translation. In *Proceedings of the 40th Annual Meeting on Association for Computational Linguistics.*

Och, F. J., Ueffing, N., and Ney, H. (2001). An efficient A* search algorithm for statistical machine translation. In *Proceedings of the Workshop on Data-driven Methods in Machine Translation-Volume 14*, pages 1–8. Association for Computational Linguistics.

Papineni, K., Roukos, S., Ward, T., and Zhu, W.-J. (2002). BLEU: A method for automatic evaluation of machine translation. In *Proceedings of the Meeting of the Association for Computational Linguistics.*

Patel, R., Gupta, R., Pimpale, P., and Sasikumar, M. (2013). Reordering rules for English-Hindi SMT. In *Proceedings of the Second Workshop on Hybrid Approaches to Translation.*

Paul, M., Finch, A., and Sumita, E. (2011). Integration of multiple bilingually-trained segmentation schemes into statistical machine translation. In *Proceedings of the Joint 5th Workshop on Statistical Machine Translation and MetricsMATR*, pages 400–408.

Paul, M., Finch, A., and Sumita, E. (2013). How to choose the best pivot language for automatic translation of low-resource languages. *ACM Transactions on Asian Language Information Processing (TALIP)*, 12(4):14.

Post, M., Callison-Burch, C., and Osborne, M. (2012). Constructing parallel corpora for six Indian languages via crowdsourcing. In *Proceedings of the Seventh Workshop on Statistical Machine Translation.*

Pouliquen, B., Steinberger, R., Ignat, C., Irina, T., and Widiger, A. (2005). Multilingual person name recognition and transliteration. *Corela*, HS-2.

Pourdamghani, N. and Knight, K. (2005). Deciphering Related Languages. In *Empirical Methods in Natural Language Processing.*

Prakash, R. (2012). Quillpad multilingual predictive transliteration system. In *Proceedings of the Second Workshop on Advances in Text Input Methods*, pages 107–113.

Rama, T. and Gali, K. (2009). Modeling machine transliteration as a phrase based statistical machine translation problem. In *Proceedings of the 2009 Named Entities Workshop: Shared Task on Transliteration*, pages 124–127. Association for Computational Linguistics.

Ramanathan, A., Hegde, J., Shah, R., Bhattacharyya, P., and Sasikumar, M. (2008). Simple Syntactic and Morphological Processing Can Help English-Hindi Statistical Machine Translation. In *International Joint Conference on Natural Language Processing*.

Ramasamy, L., Bojar, O., and Žabokrtský, Z. (2012). Morphological processing for English-Tamil statistical machine translation. In *Proceedings of the Workshop on Machine Translation and Parsing in Indian Languages*.

Ravi, S. and Knight, K. (2009). Learning phoneme mappings for transliteration without parallel data. In *Proceedings of Human Language Technologies: The 2009 Annual Conference of the North American Chapter of the Association for Computational Linguistics*.

Ravi, S. and Knight, K. (2011). Deciphering foreign language. In *Proceedings of the 49th Annual Meeting of the Association for Computational Linguistics: Human Language Technologies-Volume 1*, pages 12–21. Association for Computational Linguistics.

Rissanen, J. (1985). *Minimum description length principle*. Wiley Online Library.

Robbeets, M. I. (2005). *Is Japanese related to Korean, Tungusic, Mongolic and Turkic?* Otto Harrassowitz Verlag.

Rudramurthy, V., Khapra, M., and Bhattacharyya, P. (2016). Sharing Network Parameters for Crosslingual Named Entity Recognition. *arXiv preprint arXiv:1607.00198*.

Saha, A., Khapra, M. M., Chandar, S., Rajendran, J., and Cho, K. (2016). A Correlational Encoder Decoder Architecture for Pivot Based Sequence Generation. In *International Conference on Computational Linguistics*.

Sajjad, H., Fraser, A., and Schmid, H. (2011). An algorithm for unsupervised transliteration mining with an application to word alignment. In *Proceedings of the 49th Annual Meeting of the Association for Computational Linguistics: Human Language Technologies-Volume 1*.

Sajjad, H., Fraser, A., and Schmid, H. (2012). A statistical model for unsupervised and semi-supervised transliteration mining. In *Proceedings of the 50th Annual Meeting of the Association for Computational Linguistics*.

Samudravijaya and Murthy, H. (2012). Indian language speech sound label set. `https://www.iitm.ac.in/donlab/tts/downloads/cls/cls_v2.1.6.pdf`. [Online; accessed 9-August-2020].

Saussure, F. D. (1916). *Course in general linguistics*. Columbia University Press.

Schuster, M. and Nakajima, K. (2012). Japanese and Korean Voice Search. In *IEEE International Conference on Acoustics, Speech and Signal Processing.*

Sennrich, R., Firat, O., Cho, K., Birch, A., Haddow, B., Hitschler, J., Junczys-Dowmunt, M., Läubli, S., Barone, A. V. M., Mokry, J., et al. (2017). Nematus: A Toolkit for Neural Machine Translation. In *Software Demonstrations of the 15th Conference of the European Chapter of the Association for Computational Linguistics*.

Sennrich, R., Haddow, B., and Birch, A. (2016). Neural Machine Translation of Rare Words with Subword Units. In *Conference of the Association for Computational Linguistics*.

Sherif, T. and Kondrak, G. (2007). Substring-based transliteration. In *Annual Meeting-Association for Computational Linguistics*.

Singh, A. K. (2006). A computational phonetic model for Indian language scripts. In *Constraints on Spelling Changes: Fifth International Workshop on Writing Systems*.

Sinha, R., Sivaraman, K., Agrawal, A., Jain, R., Srivastava, R., and Jain, A. (1995). ANGLABHARTI: A Multilingual Machine Aided Translation Project on Translation from English to Indian Languages. In *IEEE International Conference on Systems, Man and Cybernetics*.

Smit, P., Virpioja, S., Grönroos, S.-A., and Kurimo, M. (2014). Morfessor 2.0: Toolkit for statistical morphological segmentation. In *Proceedings of the Demonstrations at the 14th Conference of the European Chapter of the Association for Computational Linguistics*.

Sproat, R. (2003). A formal computational analysis of indic scripts. In *International Symposium on Indic Scripts: Past and Future, Tokyo*.

Srivastava, N., Hinton, G., Krizhevsky, A., Sutskever, I., and Salakhutdinov, R. (2014). Dropout: A simple way to prevent neural networks from overfitting. *The Journal of Machine Learning Research*, 15(1), 1929–1958.

Stolcke, A. et al. (2002). SRILM-an extensible language modeling toolkit. In *INTERSPEECH*.

Subbārāo, K. V. (2012). *South Asian languages: A syntactic typology*. Cambridge University Press.

Surana, H. and Singh, A. K. (2008). A more discerning and adaptable multilingual transliteration mechanism for Indian languages. In *Proceedings of the Third International Joint Conference on Natural Language Processing*, pages 64–71.

Sutskever, I., Vinyals, O., and Le, Q. V. (2014). Sequence to Sequence learning with neural networks. In *Proceedings of Advances in Neural Information Processing Systems*.

Tao, T., Yoon, S.-Y., Fister, A., Sproat, R., and Zhai, C. (2006). Unsupervised named entity transliteration using temporal and phonetic correlation. In *Proceedings of the 2006 Conference on Empirical Methods in Natural Language Processing*.

Taskar, B., Lacoste-Julien, S., and Klein, D. (2005). A discriminative matching approach to word alignment. In *Proceedings of the conference on Human Language Technology and Empirical Methods in Natural Language Processing*.

Thomason, S. (2000). Linguistic areas and language history. In Nerbonne, J. and Schaeken, J., editors, *Languages in contact*. Editions Rodopi B.V., Brill.

Thu, Y. K., Pa, W. P., Utiyama, M., Finch, A. M., and Sumita, E. (2016). Introducing the Asian Language Treebank (ALT). In *Language Resources and Evaluation Conference*.

Tiedemann, J. (2009). Character-based PSMT for closely related languages. In *Proceedings of the 13th Conference of the European Association for Machine Translation*.

Tiedemann, J. (2012a). Character-based pivot translation for under-resourced languages and domains. In *Proceedings of the 13th Conference of the European Chapter of the Association for Computational Linguistics*.

Tiedemann, J. (2012b). Parallel Data, Tools and Interfaces in OPUS. In *Language Resources and Evaluation Conference*, volume 2012, pages 2214–2218.

Tiedemann, J. and Nakov, P. (2013). Analyzing the use of character-level translation with sparse and noisy datasets. In *Recent Advances in Natural Language Processing*.

Tillmann, C. (2004). A Unigram Orientation Model for Statistical Machine Translation. In *Annual Conference of the North American Chapter of the Association for Computational Linguistics*, pages 101–104. Association for Computational Linguistics.

Trubetzkoy, N. (1928). Proposition 16. In *Actes du premier congres international des linguistes à La Haye*.

Tyers, F. M. and Alperen, M. S. (2010). South-East European Times: A parallel corpus of Balkan languages. In *Workshop on Exploitation of multilingual resources and tools for Central and (South) Eastern European Languages*.

Utiyama, M. and Isahara, H. (2007). A Comparison of Pivot Methods for Phrase-Based Statistical Machine Translation. In *Conference of the North American Chapter of the Association for Computational Linguistics*, pages 484–491.

van der Maaten, L. and Hinton, G. (2008). Visualizing data using t-SNE. *Journal of Machine Learning Research*, 9(11), 2579–2605.

Vilar, D., Peter, J.-T., and Ney, H. (2007). Can we translate letters? In *Proceedings of the Second Workshop on Statistical Machine Translation*.

Virpioja, S. and Grönroos, S.-A. (2015). LeBLEU: N-gram-based translation evaluation score for morphologically complex languages. In *Proceedings of the Workshop on Machine Translation*.

Virpioja, S., Smit, P., Grönroos, S.-A., Kurimo, M., et al. (2013). Morfessor 2.0: Python implementation and extensions for Morfessor Baseline. Technical report, Aalto University.

Virpioja, S., Väyrynen, J. J., Creutz, M., and Sadeniemi, M. (2007). Morphology-aware statistical machine translation based on morphs induced in an unsupervised manner. In *Machine Translation Summit XI*.

Vovin, A. (2010). *Korea-Japonica: A re-evaluation of a common genetic origin*. University of Hawaii Press.

Vrandečić, D. and Krötzsch, M. (2014). Wikidata: A free collaborative knowledgebase. *Communications of the ACM*, 57(10), 78–85.

Wagner, R. and Fischer, M. J. (1974). The string-to-string correction problem. *Journal of the ACM*, 21(1), 168–173.

Wang, P., Nakov, P., and Ng, H. T. (2012). Source language adaptation for resource-poor machine translation. In *Proceedings of the 2012 Joint Conference on Empirical Methods in Natural Language Processing and Computational Natural Language Learning*.

Wang, P., Nakov, P., and Ng, H. T. (2016). Source language adaptation approaches for resource-poor machine translation. *Computational Linguistics*, 42(2), 277–306.

Williams, P., Sennrich, R., Nadejde, M., Huck, M., Haddow, B., and Bojar, O. (2016). Edinburgh's statistical machine translation systems for WMT16. In *Proceedings of the First Conference on Machine Translation*.

Wu, H. and Wang, H. (2007). Pivot language approach for phrase-based statistical machine translation. *Machine Translation*, 21(3):165–181.

Wu, H. and Wang, H. (2009). Revisiting pivot language approach for machine translation. In *Proceedings of the Joint Conference of the 47th Annual Meeting of the ACL and the 4th International Joint Conference on Natural Language Processing*.

Wu, Y., Schuster, M., Chen, Z., Le, Q. V., and Norouzi, M. (2016). Google's Neural Machine Translation System: Bridging the Gap between Human and Machine Translation. *ArXiv e-prints: abs/1609.08144.*

Xu, J., Gao, J., Toutanova, K., and Ney, H. (2008). Bayesian semi-supervised chinese word segmentation for statistical machine translation. In *Proceedings of the 22nd International Conference on Computational Linguistics*, pages 1017–1024.

Yamada, K. and Knight, K. (2001). A syntax-based statistical translation model. In *Proceedings of the 39th Annual Meeting on Association for Computational Linguistics*, pages 523–530. Association for Computational Linguistics.

Yang, Z., Salakhutdinov, R., and Cohen, W. (2016). Multi-task Cross-lingual Sequence Tagging from Scratch. *arXiv preprint arXiv:1603.06270.*

Yoon, S.-Y., Kim, K.-Y., and Sproat, R. (2007). Multilingual transliteration using feature based phonetic method. In *Annual Meeting-Association for Computational Linguistics*, page 112.

Zoph, B. and Knight, K. (2016). Multi-source Neural Translation. In *Conference of the North American Chapter of the Association for Computational Linguistics.*

Zoph, B., Yuret, D., May, J., and Knight, K. (2016). Transfer Learning for Low-resource Neural Machine Translation. In *Conference on Empirical Methods in Natural Language Processing.*

Index

Italicized and **bold** pages refer to figures and tables respectively.

Printed in the United States
by Baker & Taylor Publisher Services

Printed in the United States
by Baker & Taylor Publisher Services